Eva Interglaciation Forest Bed, Unglaciated East-Central Alaska: Global Warming 125,000 Years Ago

Goldstream Formation
Eva Forest Bed
Unconformity
Gold Hill Coess

Frontispiece. The green color of the perennially frozen Gold Hill Loess is an excellent and unique physical evidence of thawing of permafrost in loess in the Yukon-Tanana Upland during the Eva Interglaciation, 125,000 years ago. The massive pre-Sangamon Gold Hill Loess with a green color occurs in fresh exposures underlying the Eva Forest Bed and/or the Goldstream Formation. It is believed that the iron coating on the minerals (mainly quartz) in the loess was reduced from ferric (tan) to ferrous (greenish) by deeply penetrating percolating surface water as the loess was thawed under a vegetative cover in anaerobic conditions. Refreezing in Wisconsin time has preserved the green color until the loess is exposed and thawed today when it becomes a traditional tan color.

East wall of Eva Bench Mining Cut 100 m west of Eva Creek (Fig. 3), 14 km west of Fairbanks, Alaska. Wood and pipes on gravel floor of cut in foreground and on eroded surface of Goldstream Formation at left are from placer gold mining operations. White gravel at top of section is modern mine tailings. Photograph PK 1295 by T. L. Péwé, September 24, 1949.

Eva Interglaciation Forest Bed, Unglaciated East-Central Alaska: Global Warming 125,000 Years Ago

by
Troy L. Péwé
Department of Geology
Arizona State University
Tempe, Arizona 85287

Glenn W. Berger
Quaternary Sciences Center
Desert Research Institute
Reno, Nevada 89526-0220

John A. Westgate
Department of Geology
University of Toronto
Scarborough Campus
Scarborough, Ontario
M1C 1A4 Canada

Peter M. Brown*

and

Steven W. Leavitt
Laboratory of Tree-Ring Research
The University of Arizona
Tucson, Arizona 85721

SPECIAL PAPER

319

1997

*Present address: Rocky Mountain Tree-Ring Research, Inc.; 2901 Moore Lane; Fort Collins, Colorado 80526.

Published by The Geological Society of America, Inc.
3300 Penrose Place, P.O. Box 9140, Boulder, Colorado 80301

Printed in U.S.A.

GSA Books Science Editor Abhijit Basu

Library of Congress Cataloging-in-Publication Data

Eva Interglaciation Forest Bed, unglaciated east-central Alaska : global warming 125,000 years ago / by Troy L. Péwé ... [et al.].
p. cm. -- (Special paper ; 319)
Summaries in French and German.
Includes bibliographical references (p. -).
ISBN 0-8137-2319-1
1. Trees, Fossil--Alaska. 2. Geochronometry--Alaska. 3. Paleobotany--Pleistocene. 4. Glacial climates--Alaska. 5. Geology, Stratigraphic--Pleistocene. I. Péwé, Troy Lewis, 1918- . II. Series: Special papers (Geological Society of America) ; 319.
QE991.E93 1997
551.69798--dc21 97-25874
CIP

Cover photograph: Stem disk of spruce (*Picea* sp.) cut from log preserved in permafrost in Eva Forest Bed, 125,000 years old. Tree was 144 years old and grew at elevation of 250 m on south-facing slope. Collected by Péwé and Westgate in 1987, from mining exposure at Eva Creek, 14 km west of Fairbanks, Alaska, through the courtesy of Walter Wigger and Mike Wigger, mine owners. Stem disk prepared and tree rings counted by Peter Brown, Laboratory for Tree Ring Research, University of Arizona, Tucson. Péwé sample No. 187 (Table 1) Photograph No. 30,373 by S. M. Selkirk, Arizona State University, Tempe, February 20, 1997.

10 9 8 7 6 5 4 3 2 1

Contents

Geological Society of America
Special Paper 319
1997

Eva Interglaciation Forest Bed, Unglaciated East-Central Alaska: Global Warming 125,000 Years Ago

ABSTRACT

The Eva Interglaciation Forest Bed represents a frozen, buried, ancient boreal forest in the Yukon-Tanana Upland of east-central Alaska. It consists of excellently preserved peat lenses, sticks, roots, and logs as well as rooted and unrooted stumps of trees, mainly spruce and birch. Consistent with the modern boreal forest, the largest and most common tree in the fossil forest is spruce, mainly white spruce (*Picea glauca*). Remains of birch trees are common, mostly *Betula papyrifera*. The forest remains were buried by loess that became frozen and so are well preserved. None of the wood is mineralized. Many of the fragments are black from buring, suggesting forest fires were widespread in the Yukon-Tanana Upland during the interglaciation. Also, evidence is presented for the first time of the existence of spruce bark beetles (Scolytidae) during the last interglaciation in Alaska.

Efforts to determine the age of the Eva Forest Bed in this study have covered the past 50 years. Methods applied have varied from the use of stratigraphic interpretation of sedimentological events and preserved evidence of climatic changes to the use of modern geochronometry. Several methods of dating have come to fruition in the 1990s. New radiocarbon dating by liquid scintillation (LS) detectors indicates the forest wood to be older than 70,000 years. Perhaps the greatest breakthrough is the development of the isothermal-plateau fission-track method of dating geologically young volcanic glass shards. The Old Crow tephra closely underlying the Eva Forest Bed has been dated at 140 ± 10 ka and strongly supports the original interpretation of the forest bed as of last interglaciation. In the early 1990s, highly improved thermoluminescence (TL) sediment dating techniques were utilized for dating loess above and below the forest bed indicating the age of the Eva Forest Bed is probably 125,000 years with a duration of the Eva Interglaciation of probably only a few thousand years (Sangamon, Oxygen Isotope Substage 5e).

Stratigraphically, the Eva Forest Bed lies at the prominent unconformity between the underlying massive, green Gold Hill Loess (pre-Sangamon) and the overlying blackish, ice-wedge-rich retransported loess of the Goldstream Formation (Wisconsin). Studies of the frozen Gold Hill Loess indicate that the warm interglacial interval was characterized by deep and rapid thawing of permafrost and erosion of loess accompanied by gullying and block slumping of frozen loess. After extensive slumping, the topography became smooth and the forest became extensive. Tilting of enclosed tephra layers outline the slump blocks. Evidence for deep permafrost thawing is supported by the absence today of ice wedges, buried pingos, and mammal carcasses in the presently refrozen loess of pre-Wisconsin age. Deep thawing is also indicated by reduction of iron on loess grains from ferric to ferrous turning the traditional tan color of loess to greenish in the buried Gold Hill Loess. It is the unique sequence of refreezing in Wisconsin time that has preserved the remarkable evidence for deep thawing in earlier Sangamon

Péwé, T. L., Berger, G. W., Westgate, J. A., Brown, P. M., and Leavitt, S. W., 1997, Eva Interglaciation Forest Bed, Unglaciated East-Central Alaska: Global Warming 125,000 Years Ago: Boulder, Colorado, Geological Society of America Special Paper 319.

time—the green color. The forest bed formed after much of the thawing, erosion, and slumping activity had ceased, and it overlies the angular unconformity. More than half-a-dozen distinct tephra layers have been identified, characterized, and correlated in the upper part of the Gold Hill Loess, aiding in the reconstruction of the sequence of events leading to the erosion, thawing, and emplacement of the Eva Forest Bed.

Dendrochronology studies of trees and $^{13}C/^{12}C$ isotopic ratios of wood from the Eva Forest Bed, and comparisons with wood from the modern boreal forest, strongly suggest environmental conditions at least similar to those of today. Some plant remains and ground beetle taxa of Eva Forest time in Canada represent species that extended farther north than they do today. Also, buried spruce macrofossils suggest that the boreal forest may have extended north of the Brooks Range in Alaska.

These botanical and physical data indicate an environment warmer than the present interglaciation with the mean annual air temperature warmer than 0 °C, perhaps +1 or +2 °C or warmer to permit the ice to melt and permafrost to thaw from the surface downward. Supporting this concept are astronomical inferences that during the last interglacial (Oxygen Isotope Substage 5e) the July insolation anomaly at 65°N. latitude reached values of almost 50% higher than 10,000 years ago, the beginning of the Holocene Interglaciation.

ZUSAMMENFASSUNG

Das “Eva Interglaciation Forest Bed” stellt einen gefrorenen, sedimentüberdeckten, ehemaligen borealen Wald im Yukon Tanana-Hochland von Ost-Zentralalaska dar. Es besteht aus hervorragend erhaltenen Torflinsen, Ästen, Wurzeln und Baumstämmen sowie aus bewurzelten und wurzellosen Baumstümpfen, hauptsächlich von Fichten und Birken. Fichten, insbesondere Picea glauca (Weißfichte), sind die größten und häufigsten Bäume dieses fossilen Waldes, der somit mit den heutigen borealen Wäldern vergleichbar ist. Überreste von Birken, hauptsächlich von Betula papyrifera, sind häufig. Die Waldüberreste wurden durch Löß überschüttet, das anschließende Gefrieren des Materials sorgte für gute Erhaltung. Es fand keine Mineralisierung des Holzes statt. Viele der Bruchstücke zeigen Brandspuren, was die Vermutung nahelegt, daß Waldbrände während des Interglazials im Yukon Tanana-Hochland weit verbreitet waren. Erstmals wird auch der Beweis für das Vorkommen des Borkenkäfers (Scolytidae) während des letzten Interglazials in Alaska erbracht.

Die Bemühungen zur Festlegung des Alters der Eva Forest Bed reichen fünfzig Jahre zurück. Die angewendeten Arbeitsmethoden reichen von der stratigraphischen Interpretation von Sedimenterfolgen und den darin enthaltenen Hinweisen auf Klimaänderungen bis zum Einsatz moderner Geochronometrie. Verschiedene Datierungsmethoden brachten erst in den neunziger Jahren Erfolg. Neue Radiocarbon Kartierungen mittels LS-Detektoren (Liquid Scintillation) deuten darauf hin, daß der Wald älter als 70000 Jahre ist. Der vielleicht größte Durchbruch ist die Datierung der geologisch jungen vulkanischen Glasbruchstücke mittels der IPFT-Methode (Isothermal-Plateau Fisson-Track). Die das Eva Forest Bed unterlagernden Old Grow Tephra wurden auf 140000 Jahre ±10000 datiert, was sehr stark die ursprüngliche Einordnung der Forest Bed in das letzte Interglazial unterstützt. In den frühen neunziger Jahren wurden die stark verbesserten Thermolumineszenz-Sedimentdatierungsmethoden eingesetzt, um den Löß über und unter den Forest Bed zu datieren. Sie deuten auf ein Alter des Eva Forest Bed von wahrscheinlich 125000 Jahren mit einer Dauer des Eva Interglazials von wahrscheinlich nur wenigen tausend Jahren (Sangamon, Oxygen Isotope Substage 5e).

Die Eva Forest Bed liegen an der bekannten stratigraphischen Grenze zwischen dem unterlagernden, grünen Gold Hill Loess (pre-Sangamon) und den darüber liegenden schwarzen, eiskeilreichen, umgelagerten Lössen der Goldstream Formation (Wisconsin). Studien des gefrorenen Gold Hill Loess zeigen an, daß die warme interglaziale Phase durch ein tiefes und rasches Auftauen des Permafrostes und durch Erosion von Löß charakterisiert ist, begleitet durch Gullybildung und Sackungen von gefrorenem Löß. Infolge der ausgedehnten Materialverlagerungen wurde die Topographie ausgeglichen und der Wald konnte sich stark ausbreiten. Die Rutschungen werden durch die schräg gestellten Tephralagen charakterisiert. Das tiefe Auftauen des Permafrostes wird durch das Fehlen von Eiskeilen, überschütteten Pingos, Kadavern von Säugetieren in der heute wiederum gefroreren Löß präwisconsinzeitlichen Alters belegt. Auch die Reduktion von Eisen auf Lößpartikel belegt das tiefe Auftauen des Permafrostes, der überschüttete Gold Hill-Löss wird dadurch grünlich. Die grüne Farbe als bemerkenswerter Beweis für das tiefe Auftauen in der frühen Sangamonzeit verdankt ihre Erhaltung dem einzigartigen Wiedergefrieren im Wisconsin. Die Forest Bed entwickelten sich nach Beendigung der Auftau-, Erosions- und Rutschaktivität. Über ein halbes Dutzend verschiedener Tephralagen im oberen Teil des Gold Hill Lösses wurden identifiziert, charakterisiert und miteinander verglichen und halfen bei der Rekonstruktion der Ereignisabfolge, die zu Erosion, Auftauen und Ablagerung der Eva Forest Bed führten.

Die dendrochronologischen Studien und die C13, C12-Isotopenverhältnisse im Holz der Eva Forest Bed deuten stark darauf hin, daß die Umweltbedingungen zumindest ähnlich den heutigen waren. Einige pflanzliche Überreste und Laufkäfer (Taxa) aus der Eva Forest-Zeit belegen die Existenz von Arten, die heute soweit nördlich nicht mehr vorkommen. So lassen z.B. fossile Fichtenteile vermuten, daß sich der boreale Wald bis nördlich der Brooks Range in Alaska ausgedehnt hat. Diese botanischen und physikalischen Daten belegen wärmere Klimaverhältnisse als im gegenwärtigen Interglazial, die mit einer mittleren jährlichen Lufttemperatur von über 0 °C (möglicherweise 1 oder 2 °C oder mehr) das Auftauen des Eises und des Permafrostes erlaubten. Astronomische Schlußfolgerungen stützen ebenfalls diese Annahme, in dem während des letzten Interglazials die Einstrahlungsanomalien im Juli auf 65°N Werte erreichten, die um 50% über den zu Beginn der Nacheiszeit (vor 10000 Jahren) herrschenden Werte lagen.

RÉSUMÉ

La couche forestière interglaciaire Eva représente une ancienne forêt boréale enterrée et gelée dans les hautes terres drainées par le Yukon et la Tanana au centre est de l'Alaska. Elle consiste dans des lentilles de tourbe, des branches, des racines et des troncs extrêmement bien préservés, ainsi que des souches enracinées et non enracinées, principalement de sapins et de bouleaux. Comme dans les forêts boréales actuelles, les arbres les plus grands et les plus communs de cette forêt fossile étaient le sapin, et principalement le sapin blanc (*picea glauca*). Des restes de troncs de bouleaux sont communs, principalement de *Betula papyrifera*. Les restes de cette forêt ont été enterrés sous du loess et gelés, c'est pourquoi ils sont aussi bien préservés. Aucun des restes de bois n'est minéralisé. De nombreux fragments sont noircis par le feu, suggérant que les feux de forêts étaient largement répandus dans cette région pendant cet interglaciaire. En outre, pour la première fois, on peut démontrer ici l'existence de coléoptères attaquant l'écorce du sapin (scolytidae) en Alaska, pendant le dernier interglaciaire.

De nombreux efforts pour déterminer l'âge de ce lit forestier Eva ont été poursuivis pendant les 50 dernières années. Les méthodes utilisées ont couvert une large gamme, allant de l'interprétation systématique des événements sédimentologiques et d'évidences de changements climatiques, jusqu'à l'emploi de géochronométres modernes. Plusieurs méthodes de datation ont porté des fruits dans les années 1990. Des datations par C14 effectuées par scintillation liquide indiquent que les bois trouvés sont plus vieux que 70000 ans. Mais la plus grande avancée résulte d'une méthode de détermination des traces de fission utilisée habituellement pour dater des éclats de verre volcaniques. Le tephra "Old Crow," un peu inférieur à la couche Eva étudiée, a été daté par cette méthode de 140 ± 10 KA et étaye très fort l'interprétation interglaciaire de cette couche. Au début des années 1990, des techniques de thermoluminescence hautement améliorées ont été utilisées pour dater le loess au-dessus et en-dessous de la couche Eva. Elles ont indiqué que l'âge probable de cette couche est de 125000 ans et que la durée de formation de la couche elle-même a été probablement seulement de quelques milliers d'années (Sangamon: sous-stage isotopique 5e).

Stratigraphiquement, la couche forestière Eva se trouve immédiatement au-dessus de la grande discordance qui existe entre le loess massif sous-jacent dénomme "Gold Hill" (pre-Sangamon) et le loess noirâtre riche en glace de fentes de gel des formations de "Goldstream" (Wisconsin). Des études du loess gelé "Gold Hill" indiquent que l'intervalle chaud interglaciaire était caractérisé par un profond et rapide dégel du pergélisol et une érosion du loess accompagnée par des ravinements et des glissements de blocs de loess gelés. Après ces glissements très nombreux, la topographie devint douce et la forêt a couvert le territoire. L'inclinaison de couches de tephra démontre la réalité des glissements rotationnels des blocs de loess. Le dégel profond du pergélisol est aussi indiqué par la réduction du fer du loess, le transformant de fer ferrique en fer ferreux et remplaçant la couleur jaune traditionnelle des loess en la couleur grisâtre du loess enterré "Gold Hill." C'est parce qu'il n'y a eu qu'un seul gel au Wisconsinien que la trace (la couleur verdâtre) d'un dégel profond au début de la période Sangamon a été conservée. La couche Eva formée après que le dégel, l'érosion et les activités de glissement de terrain se soient terminées, est au-dessus de la discordance. Plus d'une demi-douzaine de couches distinctes de tephra ont été identifiées, caractérisées et corrélées au sein de la partie supérieure du loess "Gold Hill." Elles ont aidé à la reconstruction de la séquence d'événements conduisant à l'érosion, au dégel et à l'emplacement de la couche forestière Eva.

Les études dendrochronologiques des arbres et les pourcentages isotopiques $^{13}C/^{12}C$ de bois de cette couche ainsi que des comparaisons avec du bois de la forêt boréale actuelle, suggèrent fortement des conditions environnementales semblables à celles que nous connaissons aujourd'hui. Quelques groupes de plantes et de coléoptères du sol de cette couche forestière représentent des espèces qui s'étendaient alors plus au nord que maintenant. Par exemple des macrofossiles de sapins suggèrent que la forêt boréale s'est étendue au nord de la chaîne de Brooks en Alaska.

Ces données botaniques et physiques indiquent un environnement plus chaud que le périglaciaire actuel avec une température moyenne annuelle plus chaude que 0 °C, et peut-être même atteignant +1 à +2 °C, ce qui a permis à la glace de fondre et au pergélisol de dégeler depuis la surface vers la profondeur. Les déductions provenant des observations astronomiques parlent en faveur de ce interprétation, car pendant le dernier interglaciaire (sous-stage isotopique 5e), l'anomalie d'insolation de juillet, à 65° de latitude nord, atteignait des valeurs presque 50% plus élevées qu'il y a 10000 ans, au début de l'Holocène.

РЕФЕРАТ

Стратиграфический комплекс Forest Bed межледниковья Eva представляет собой промерзший и погребенный, древний северный лес, находящийся в пределах Юконо-Тананского нагорья восточно-центральной Аляски. Этот комплекс состоит из прекрасно сохранившихся линз торфа, кусочков веток, корней и бревен, а также пней деревьев с корнями и без них, представленных главным образом елью и березой. В согласии с ситуацией в современном северном лесу, наиболее крупное и часто встречаемое дерево в ископаемом лесу в основном представлено белой елью (Picea glauca). Останки березовых деревьев, которые наиболее распространены, представлены главным образом Betula papyrifera. Останки этого лесса были погребены под отложениями леса, подвергшегося промерзанию, и таким образом оказались хорошо сохраненными. Ни один образец древесины не был минерализован. Многие из фрагментов древесины потемнели в результате обугливания, подтверждая тем самым предположение о том, что в течение рассматриваемого межледниковья пожары в пределах Юконо-Тананского нагорья были широко распространенным явлением. Также существует свидетельство в пользу первоначального появления еловых жучков-короедов (Scolytidae) в течение последнего межледниковья на Аляске.

Усилия, направленные на определение возраста комплекса Eva Forest Bed в данном иследовании, охватывают последние 50 лет. Методы, применявшиеся для этого, колебались от стратиграфической интерпретации седиментологических событий и сохранившихся свидетельств климатических измерений до использования современной геохронометрии. Несколько методов датировки появилось в девяностые годы. Использование радиоуглеродных датировок с помощью детекторов стинциляции жидкости (LS) показало, что древесина погребенного леса должна быть старше 70.000 лет. Наибольший прорыв был достигнут при развитии трекового метода датировки геологически молодых обломков вулканических стекол. Пирокластические отложения Old Crow, подстилающие комплекс отложений Eva Forest Bed, датированы возрастом 140±10 тыс. лет, что сильно поддерживает интерпретацию происхождения залежей леса как результат последнего межледниковья. В начале девяностых годов, для датировки лессов, залегающих выше и ниже лесного комплекса, была использована улучшенная технология термолюминисцентной датировки осадков (TL), показавшая что возраст комплекса Eva Forest Bed составляет вероятно 125.000 лет с продолжительностью межледниковья Eva вероятно только в несколько тысяч лет (Сангамон, изотоп кислорода подстадии 5e).

Комплекс Eva Forest Bed стратиграфически залегает на четко выраженном несогласном залегании подстилающего массивного, зеленого цвета лесса Gold Hill Loess (пред-Сангамон) и перекрывающего черноватого лесса формации Goldstream Formation, переотложенного и обогащенного ледяными клиньями (Висконсин). Результаты исследования мерзлого лесса Gold Hill Loess показывают, что теплый межледниковый период характеризовался глубоким и быстрым оттаиванием мерзлоты и эрозией лесса, сопровождавшихся образованием оврагов и оползанием мерзлого лесса. После обширного процесса оползания лесса, поверхность выравнилась и область распространения леса увеличилась. Наклонение вмещающихся слоев пирокластического материала очерчивает блоки оползания. Свидетельство в пользу глубокого оттаивания мерзлоты подтверждается тем, что в лессе пред-Висконсинского возраста (проморoженном в настоящий период времени) отсутствуют ледяные клинья, погребенные пинго и скелеты мамонтов. Глубокое оттаивание также подтверждается изменением формы железа на зернах лесса от трехвалентного до двухвалентного, что привело к изменению цвета лесса от традиционного рыжевато-коричневого до зеленоватого в пределах погребенного лесса Gold Hill Loess. Это представляет собой уникальную последовательность повторного промораживания во время Висконсина, которое сохранило замечательное свидетельство в пользу глубокого оттаивания Сангамонского времени -- зеленый цвет. Залежи леса, сформированного после того как прекратилась деятельность наибольшего оттаивания, эрозии и оползания, перекрывают угловое несогласие залегания осадков. В верхней части лесса Gold Hill Loess были идентифицированы, охарактеризованы и скоррелированы более чем пол дюжины отчетливых слоев пирокластических отложений, что помогло провести реконструкцию последовательности событий, приведших к эрозии, оттаиванию и переотложению комплекса осадков Eva Forest Bed.

Результаты изучения дендрохронологии деревьев и изотопных отношений $^{13}C/^{12}C$ в древесных остатках из комплекса Eva Forest Bed и сравнение их с таковыми по древесине из современного северного леса позволяют уверенно предположить, что условия окружающей среды того периода были, по крайней мере, близки к существующим в настоящее время. Некоторые растительные остатки и таксономические группы почвенных жуков времени Eva Forest представляют собой разновидности, которые были распространены намного севернее чем в настоящий период времени. Например, макроостанки погребенной ели позволяют предположить, что северный лес того времени мог распространяться к северу от горной цепи Brooks на Аляске.

Эти ботанические и физические данные указывают на более теплые условия окружающей среды того времени (по сравнению с современным межледниковьем) при средней годовой температуре воздуха выше 0 C, возможно +1 или +2 C или выше, чтобы обеспечить условия для плавления льда и протаивания мерзлоты от поверхности вглубь. Подтверждением этой концепции служат астрономические заключения о том, что в течение последнего межледниковья (изотоп кислорода подстадии 5e), июльская аномалия солнечной радиации на широте 65 N достигала величины почти на 50% выше, чем 10.000 лет тому назад, в начале Голоценового межледниковья.

摘　　要

阿拉斯加中东部育空-塔纳诺高地的间冰期伊娃 林层是被埋藏、冻结的北方古森林代表，它由保存完好的透镜状泥炭层、树干、树根、木块以及有根和无根的树桩组成，主要是云杉和白桦。在该化石森林中最大和最常见的树种是云杉，尤其是白云杉(*Picea glauca*) 。桦树的残体也常见，主要有*Betula papyrifera*，与现代北方森林一致。森林残体被黄土埋藏，冰冻后得以完好保存，木头无一矿化。许多碎木块由于燃烧而变成黑色，暗示在间冰期育空-塔纳诺高地发生过大面积的森林火灾。同时也提出了在阿拉斯加末次间冰期云杉棘胫小蠹科甲虫(*Scolytidae*) 首次出现的时间证据。

确定伊娃林层年代的研究持续进行了50年。所用的方法业已改变：从以地层学方法解释沉积事件和保存的气候变化证据变为用现代地质年代学手段进行年代测定。几种定年方法在90年代得到使用。以液体闪烁(LS) 探测器进行放射性碳测年指出林木老于70,000年。最伟大的突破也许是利用等温坪裂变径迹方法对年轻的火山玻璃碎片进行地质测年。下伏于伊娃林层的老乌鸦火山灰已测定为140± 10ka，有力地支持了林层形成于末次冰期的最初解释。在90年代早期，大大改进的热释光(TL) 沉积测年技术应用于该林层上下黄土的定年，指出伊娃林层的年龄可能为125,000年，即可能形成于仅持续几千年的伊娃间冰期(桑加蒙，氧同位素5e亚阶) 。

在地层上，伊娃林层显著地不整合于下伏巨厚的绿色金山黄土(前桑加蒙) 和上覆浅黑富有冰楔的再搬运金河黄土 层(威斯康辛) 之间。对金山黄土的研究指出，它具有永久冻土在深部快速溶解的特征和被侵蚀并伴随冻结黄土沟谷化和大块塌陷的特征。在大面积的崩落之后，地形变得平坦，森林的范围得以扩大。其中火山灰层的倾斜勾画出塌落的块体。其缺乏现今所见的冻土冰楔、并在现今再冻结的前威斯康辛黄土中发现了埋藏冰核丘和哺乳动物尸体，说明曾经历过永久冻土深部融化。被埋藏的金河黄土颗粒中铁离子从三价减少到二价，黄土改变了其传统的棕黄色而成为浅绿色，这也指出曾有过深部溶解。正是这些在威斯康辛时期的再冻结的特殊序列保存了较早威斯康辛时期的深部溶解的证据 — 绿色。林层形成于溶解、侵蚀和崩塌活动停止很久之后，以角度不整合盖在其上。已识别出六个以上清晰的火山灰层，其特征已确定与金山黄土的上部有联系。这有助于重建导致伊娃林层侵蚀、溶解和定位的事件序列。

伊娃林层的树轮年代学和树木$^{13}C/^{12}C$同位素比率的研究，以及与现代北方森林树木的比较，强调当时的环境状态至少和现在相似。伊娃森林时期的一些植物残体和地下甲虫类代表了比现今延伸更北的种群。比如，埋藏的云杉大化石暗示北方森林也许已延伸到阿拉斯加的布鲁克斯山脉北部。

这些植物和自然地理的信息指出那时环境比现今更温暖，年均温高于0℃，也许1到2℃或者更暖，从而导致冰层融化、永冻层自上而下溶解。这个观点也得到天文推理支持，即在末次间冰期(氧同位素5e亚阶) ，在北纬65度7月份日照率异常，其数值几乎比10,000年前即全新世间冰期开始时高50%。

INTRODUCTION

The Eva Forest Bed lies in the upper part of the thick, perennially frozen late Cenozoic loess deposits that occur throughout unglaciated east-central Alaska (Fig. 1). We show here that the forest bed provides clear evidence for a great, widespread climatic warming of 125,000 years ago, here formally named the Eva Interglaciation. In addition to the forest bed, evidence for this conclusion consists of prominent breaks in the sedimentary, climatic, floral, permafrost, and geomorphic records. Rapid thawing of the permafrost occurred with formation of widespread thermokarst terrain accompanied by a major erosional interval with gullying of loess on hillsides and removal of loess in creek valley bottoms. The thawing and erosion were accompanied and/or followed by a return of the boreal forest, the taiga, after an earlier colder, treeless glacial time. This interglacial forest is now represented by the frozen buried Eva Forest Bed of well-preserved rooted stumps and prostrate logs.

This prominant stratigraphic break in the geological record was recognized in central Alaska at Eva Creek near Fairbanks by Péwé on June 28, 1949, and reported[1] informally as the Eva Interglaciation: the latest interglaciation—the Sangamon (Oxygen Isotope Substage 5e). However, only now can this interglacial event recorded in the loess of the subarctic be more quantitatively demonstrated to be equivalent to the last interglaciation. Westgate and his associates (Westgate, 1988, Westgate et al., 1990) characterized and dated tephra in the loess and determined the paleomagnetism. In addition, Berger dated the loess near the forest bed by TL (Berger and Péwé, 1994), and Long and other scientists dated wood from the Eva Forest Bed by specialized radiocarbon dating methods (Long and Kalin, 1992a, 1992b and McCormac et al., 1993). Also, scientists at the University of Arizona Laboratory of Tree Ring Research have recently demonstrated by dendrochronological studies (Brown) and by $^{13}C/^{12}C$ isotopic-ratio studies (Leavitt) the similarity of wood from the Eva Forest Bed to the wood from modern trees.

The data summarized here include results from examination of hundreds of loess exposures exhibited when gold mining excavations were made and the loess faces were still frozen. These observations are combined with old and new mechanical, mineralogical, and chemical analyses of the loess along with data on permafrost, and with the new information mentioned in the preceding paragraph. Therefore, we present here in a historical manner, with the latest quantitative results, the study of the Eva Forest Bed over the past 50 years. We take this approach because of (1) the long-developing knowledge of both the age and paleoenvironmental character of the Eva Forest Bed; (2) the importance of all continental sedimentological records of the last interglaciation; (3) the early creation and subsequent loss by placer gold

---

[1]Péwé, 1952a, 1957, 1958, 1965b, 1965c, 1975a, 1975b, 1989, 1992, in press; Péwé and Hopkins, 1967; Péwé and Sellmann, 1973; Péwé and Reger, 1983, 1989; Péwé et al., 1989; Berger and Péwé, 1994; Berger et al., 1996; Matthews, 1970; Wintle and Westgate, 1986; Westgate, 1988.

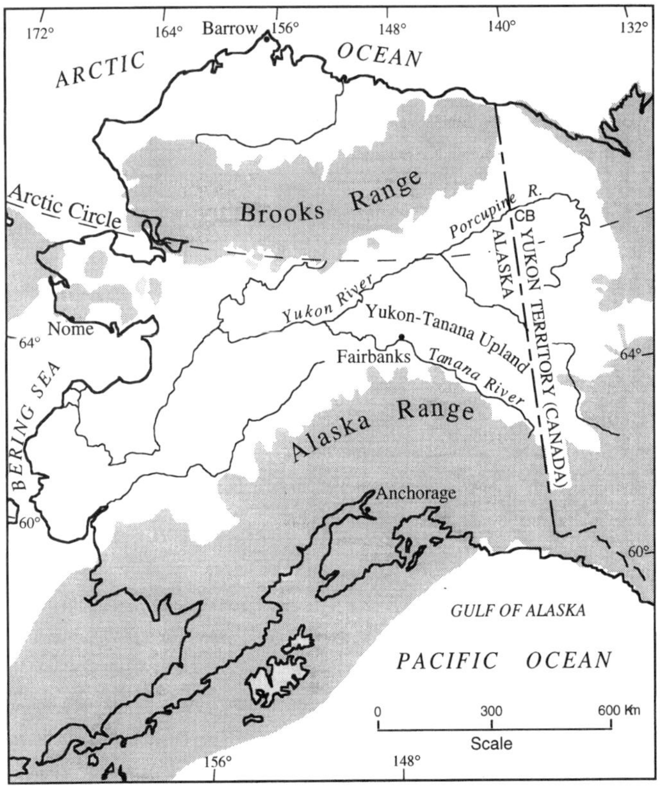

Figure 1. Extent of late Cenozoic glaciations (shaded) in Alaska and northwestern Canada and location of Yukon-Tanana Upland. From Péwé (1975a) and Tarnocai and Schweger (1991). CB indicates location of Ch'ijee's Bluff in northern Yukon Territory.

mining of the most dramatic exposures of the Eva Forest Bed, and the consequent unique documentation of descriptions of these exposures in the decades-long field notes of Péwé; and (4) the consequent importance for posterity of this otherwise inaccessible information and of this historical development of knowledge of this major Alaskan feature. This summary, honed to mesh with modern and conclusive quantitative data, will probably become increasingly valuable to Quaternary scientists (and present and future students of this region and topic) as the permafrost continues to thaw, and as the sedimentological record thereby continues to fragment and disperse spatially. We hope that our summary is both timely and useful to future Quaternarists who wish to reconstruct the geological record of the last interglaciation in central Alaska.

These long-term studies reveal that a unique set of physical events took place during this prominent interglaciation in a region characterized by a perennially frozen ice-rich blanket of easily erodible loess overlying a hilly bedrock topography of the Yukon-Tanana Upland (Fig. 2) (Péwé, in press). Extensive piping, gullying, and block slumping of frozen loess took place over a large region in the warmer-than-present erosional episode, and most permafrost disappeared.

A boreal forest of mainly spruce, birch, and poplar species developed over much of the landscape in the lower elevations of central Alaska as the climate warmed from the earlier treeless Mammoth Steppe environment (Guthrie, 1990). The forest existed for perhaps a few thousand years and then disappeared with increased loess deposition and the onset of the colder, drier glacial climate of Wisconsin time. Remnants of the perennially frozen forest bed are today sandwiched between the overlying loess of Wisconsin age and the underlying older loess.

ACKNOWLEDGMENTS

In collecting the field data and assembling this manuscript, Troy L. Péwé has drawn heavily on his experience with colleagues of the Branch of Alaskan Geology, U.S. Geological Survey and the Departments of Geology at the University of Alaska Fairbanks and at Arizona State University over the past 50 years. In 1987 Malcolm K. Hughes, Director, Laboratory of Tree-Ring Research, University of Arizona, Tucson, initiated informative discussions with personnel of the laboratory concerning study of the Alaskan wood.

In the 1940s and 1950s Otto Wm Geist of the University of Alaska introduced Péwé and his wife and field associate, Mary Jean Péwé, to the great number of "mining cuts," large and small, and to the more than a hundred industrious gold miners from one end of the Yukon-Tanana Upland to the other, as well as to those in the Dawson area in adjacent Yukon Territory, Canada. These cooperative miners generously provided access to mining exposures and frozen ground data to the Péwé's and their associates throughout the last half of the twentieth century. Especially cooperative were Roy Earling, John E. Metcalfe, and James D. Crawford, administrators of the U.S. Smelting, Refining, and Mining Company; Pete Eagan, Manager, Fairbanks Department, Alaska Gold Company; and Walter and Mike Wigger, owners of properties in the Eva Creek area.

In the field Péwé was aided by many assistants of the U.S. Geological Survey and University of Alaska, notably E. S. King, Jr., A. M. Gooding, D. R. Loftus, George Herman, E. W. Marshall, D. D. Smith, R. A. Paige, L. R. Mayo, R. D. Reger, P. V. Sellmann, N. W. Rutter, J. M. Blackwell, N. R. Rivard, J. W. Bell, and E. J. Bell. The following provided valuable field assistance to J. A. Westgate: S. J. Preece, B. Stemper, Qiang Hu, and A. Westgate. Berger thanks his co-authors and S. J. Preece for field assistance.

Special thanks go to former Alaska State Senators Bettye Fahrenkamp and Shirley Craft for obtaining financial support in 1989 to create the Gold Hill Heritage Loess Preserve at Gold Hill near Fairbanks. Loess stratigraphy at this preserve was studied for this report. From the early 1950s, E. H. Beistline, former dean of the School of Mines at the University of Alaska, provided information on mining activities and personnel. G. E. Weller and G. D. Guthrie, also of the University of Alaska, kindly provided climatic and vertebrate paleontological information respectively. F. R. Weber and J. D. Townshend of the U.S. Geological Survey provided both Péwé and Westgate logistic aid in the field over several summers and James and Sally Murphy of Fairbanks pro-

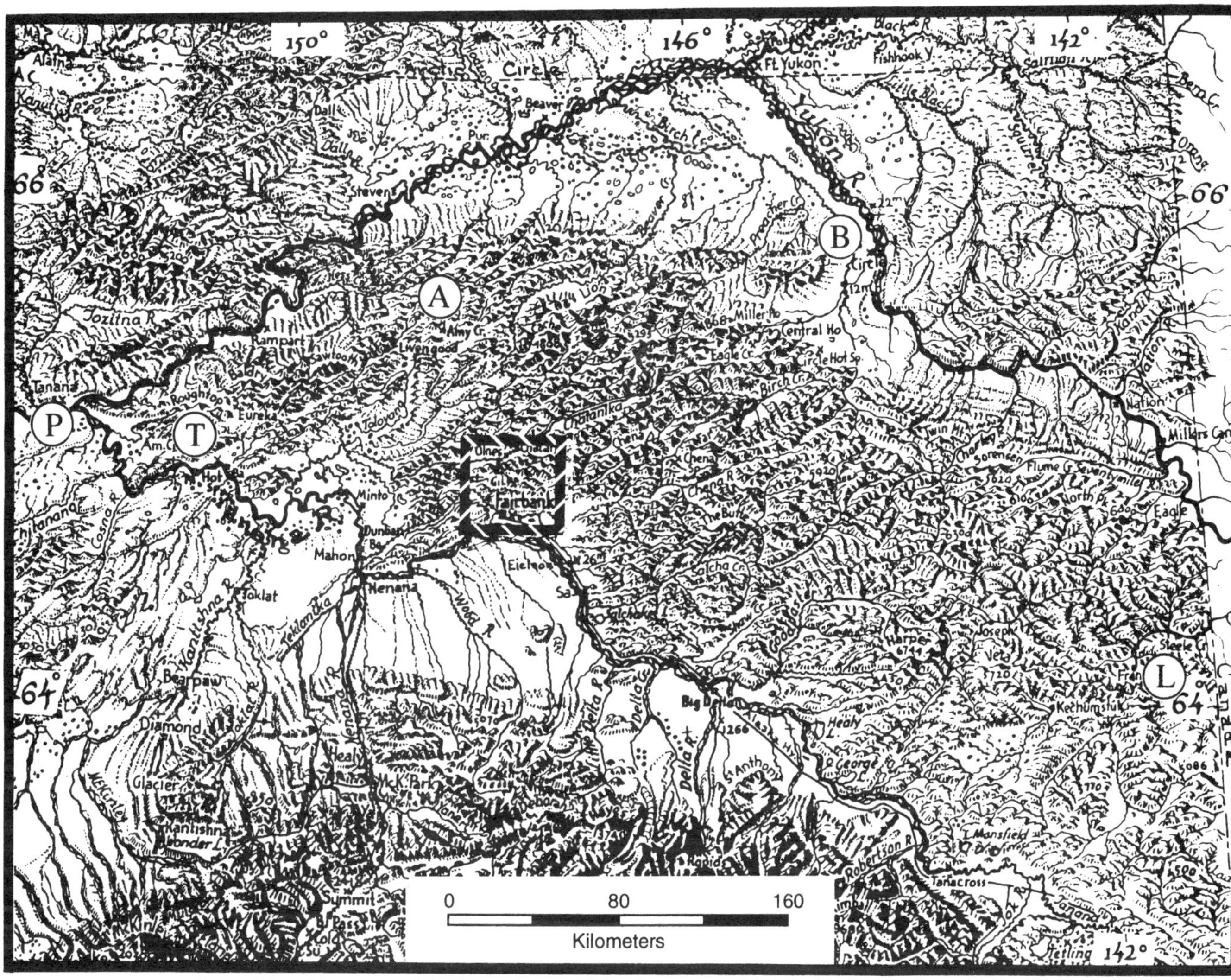

Figure 2. Landform map of the Yukon-Tanana Upland, east-central Alaska. The Fairbanks area (Fig. 3), where most of the Eva Forest Bed localities occur, is outlined by the black rectangle. White circles with letters on this map are Eva Forest Bed localities or Quaternary sections outside of the Fairbanks area: P—Palisades; T—Tofty Placer District; A—Amy Creek; B—Birch Creek site; L—Lost Chicken Creek. (Modified slightly from Erwin Raisz's Landform Map of Alaska, prepared for the Quartermaster Corps, U.S. Army, 1948.)

vided critical logistic support to Péwé during the summers of the 1970s, 1980s, and 1990s.

Special recognition is due to Richard D. Reger, formerly of the Alaska State Division of Geological and Geophysical Surveys, Fairbanks. He was associated with various aspects of the project from 1954 to 1994, aided in field work, and argued successfully, as well as unsuccessfully, scientific ideas in the field and in the office. We are grateful for his stimulating scientific discussions over these many years. Also, the authors are deeply appreciative of the thorough reviews of early drafts of the entire manuscript by Richard D. Reger; Oscar J. Ferrians, U.S. Geological Survey; and by Leslie A. Viereck, principal plant ecologist, Institute of Northern Forestry, U.S. Forest Service, Fairbanks. Three anonymous reviewers greatly improved the manuscript. The following scientists kindly translated the abstract into foreign languages: Albert Pissart; Lorenz King and Elisabeth Schmitt; Liu Tung Sheng and Tan Ming; and Ruslan O. Kuzmin and Sofiya B. Vasina.

Early drafts of the index maps, climatic charts, and Figure 4, were prepared by Mary Jean Péwé and later updated by Susan Mary Selkirk, graphic artist, Arizona State University. The figure of climatic data was compiled by Mike "Colonel" Woodhouse of Arizona State University. Except as noted, the line drawings were prepared by David A. Keadle, with early drafts of some prepared by Keenan Murray. Photographs of stem disks of modern and fossil wood are by Dan H. Ball, Arizona State University.

Fieldwork in the 1940s and 1950s was with the Alaska Terrain and Permafrost Section of the U.S. Geological Survey, supported in part by the Engineer Intelligence Division, Office of the Chief of Engineers, United States Army. In 1977, stratigraphic investigations by Péwé were supported by the State of Alaska Division of Geological and Geophysical Surveys. Financial support from the U.S. National Academy of Sciences permitted field checking by Péwé in 1981, 1982, and 1983 in connection with preparations of the Fourth International Conference on Permafrost at Fairbanks.

From 1985 to 1995, Péwé, Westgate, and Berger received grants from national agencies: Péwé-Westgate, U.S. National Science Foundation Grant EAR 8520143 (1985–1989) and EAR 9022590 (1991–1995); National Geographic grants 4100-89 and 5196-94 to Péwé for trenching and radiocarbon dating respectively; operational support for Westgate's tephra studies from the Natural Sciences and Engineering Research Council of Canada and the American Chemical Society; field laboratory support to Berger from National Science Foundation Grant EAR 9018602-001; and laboratory support by the Desert Research Institute.

Peter Brown thanks James Burns from the Laboratory of Tree-Ring Research, University of Arizona, Tucson, for X-ray densitometry of the Eva samples and Bruce Bauer from the International Tree-Ring Data Bank, National Geophysical Data Center, Boulder, Colorado, for providing the raw density data from Fritz Schweingruber's white spruce sites in Alaska.

Steven Leavitt thanks S. Danzer for helping prepare samples and T. Newberry and B. McCaleb for helping to analyze the samples. Samples of CO_2 prepared from combustion of the tree-ring cellulose were analyzed in the Laboratory of Isotope Geochemistry, Department of Geosciences, University of Arizona (Austin Long, Director).

PHYSICAL SETTING

Topography and geology

Central Alaska is regarded as the area that lies between the Brooks Range on the north and the towering Alaska Range on the south (Fig. 1). The area was not glaciated, except in small local mountainous areas, especially in the Yukon-Tanana Upland of east-central Alaska (Péwé et al., 1967; Weber, 1986), but glaciers from the Brooks Range on the north, the Alaska Range on the south (Coulter et al., 1965), and the Yukon Plateau on the east (Bostock, 1948, 1966) bordered the interior of Alaska during times of glacial maxima (Fig. 1). Indeed, glaciers from the Alaska Range probably approached within 80 km of the Fairbanks area.

The Yukon-Tanana Upland is an east-trending upland between the Yukon and Tanana Rivers (Fig. 2) and consists of a dissected area of accordant rounded ridges 700 to 1,000 m in altitude. Scattered, discontinuous groups of higher mountains project above the upland ridges to altitudes of 2,000 m. Rounded upland ridges near Fairbanks have summits at 500 to 600 m above sea level. The bedrock of the southern part of the upland is chiefly schist, slate, and gneiss, but includes local masses of basalt, quartz, diorite, and granite (Péwé et al., 1966).

During glacial advances, sediment-laden glacial rivers deposited hundreds of meters of sand and gravel in the Tanana and Yukon Valleys. Aggradation of these trunk valleys raised base level and caused tributaries from the unglaciated Yukon-Tanana Upland to aggrade their lower valleys. Especially during times of glacial expansion, winds blowing across outwash plains and vegetation-free bars of glacier-derived braided streams in trunk valleys picked up great quantities of silt and redeposited it as loess (Péwé, 1951). It was deposited on the Yukon-Tanana Upland and other nearby areas, blanketing ridges with thicknesses from a few cm to more than 60 m near rivers. Most of the loess was retransported to creek-valley bottoms, where the thickest recorded is 100 m in Isabella Creek valley near Fairbanks (Fig. 3; Péwé, 1958, drill hole no. 103).

Today the upper slopes of the bedrock hills are covered with a few cm of loess, but the middle slopes are blanketed by 3 to 20 m of loess that is dissected by old, subdued parallel gullies. These gullies are 10 to 15 m deep and 100 to 200 m long; interfluves and gully bottoms are rounded and subdued. Loess ridges are locally separated into isolated, elongate and round knobs toward their lower end. The silt-blanketed hills in the Fairbanks area are not undergoing active gullying today, except where human activity weakens the vegetative cover or concentrates runoff. The rounded gullies and ridges are relics of an ancient period of gully formation (Péwé, 1955, p. 714).

Summary of late Cenozoic stratigraphy

To illustrate the stratigraphic position of the Eva Forest Bed, a brief summary of the late Cenozoic deposits in east-central Alaska is presented (Fig. 4). The oldest unconsolidated deposit is the Cripple Gravel (Péwé, 1975a, 1989), a gold-bearing, brownish, coarse sandy gravel with angular clasts that is preserved on buried bedrock benches. The Cripple Gravel is late Tertiary in age (Westgate et al., 1990; Péwé, in press) with no reported fossils. It is interpreted as being composed of solifluction debris that was produced in a periglacial climate and partly reworked by streams. The gravel formed when drainage directions were different from those of today (Péwé 1965c).

After modern drainages had been established by rejuvenated streams, a gold-bearing, poorly stratified gravel named the Fox Gravel (Péwé, 1975a) accumulated in valley bottoms under periglacial conditions. The upper part of the Fox Gravel contains large bones of mammoth, horse, and bison that are thought to be of late Pliocene age. At various locations, the Fox Gravel is overlain by the Dawson Cut Formation, Gold Hill Loess, or Goldstream Formation. The Tanana Formation is a widespread inactive solifluction layer 1–25 m thick consisting mainly of schist and quartz fragments in a sandy matrix. It is overlain by Gold Hill Loess (Fig. 4).

The Dawson Cut Formation (forest bed) of the Dawson Cut Interglaciation (Péwé, 1952a, 1965c, 1975b, 1989, in press; Péwé et al., 1995) is a gray-to-black silt unit 1 to 3 m thick that contains peat lenses, logs, and forest beds and lies at the base of the Gold

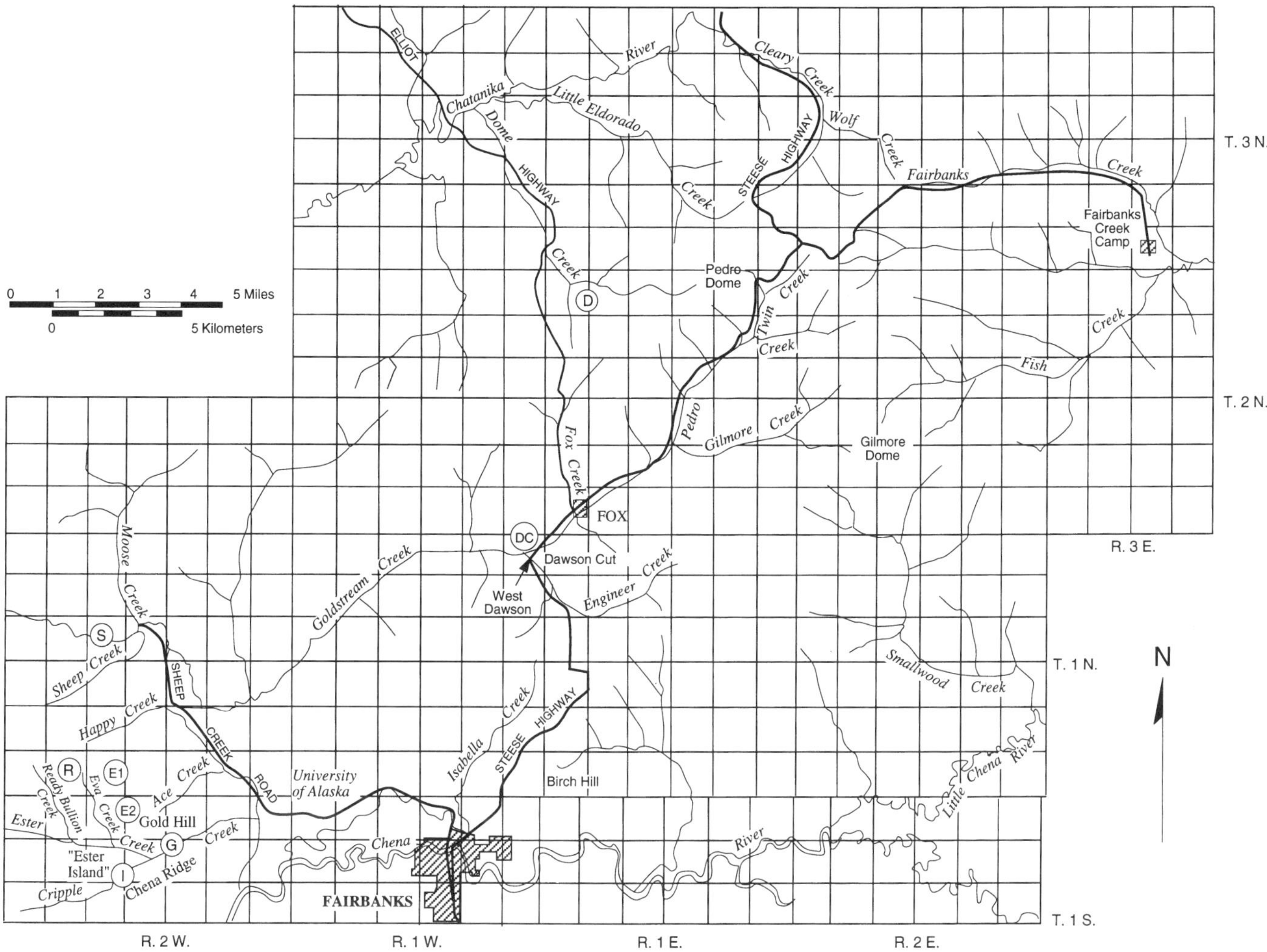

Figure 3. Index map of the Fairbanks area, south-central Yukon-Tanana Upland, Alaska, showing Eva Forest Bed localities discussed (white circles) and other geographical features. D—Dome Creek; DC—Dawson Cut; E1—Upper Eva Creek; E2—Lower Eva Creek; G—Gold Hill; I—Ester Island; R—Ready Bullion Creek; S—Sheep Creek Cut.

Hill Loess. White spruce logs are as much as 30 cm in diameter. Spruce, birch, and other species have been identified. Péwé believes that this forest bed is about 2 Ma old (Péwé et al., 1995).

The Gold Hill Loess is a thick, massive, frozen, tan to grayish, mainly airfall loess with no ice wedges. On lower slopes it locally underlies, with a marked unconformity, either the prominent, frozen, well-preserved interglaciation Eva Forest Bed or the Goldstream Formation (Péwé, 1965c, 1975b; Fig. 4). The term Gold Hill Loess was used informally in the 1950s and 1960s and was later formalized (Péwé, 1975b). On upper slopes where it could not be differentiated from loess of Wisconsin or Holocene age, it was lumped under the general term of Fairbanks Loess (Péwé, 1958, 1975b). With the identification and correlation of various tephra layers in the Gold Hill Loess, it is now possible to recognize Gold Hill Loess on upper slopes where it appears near or at the surface (Preece, 1991; Péwé, 1992, in press; Fig. 4). The term Gold Hill Loess embraces air fall and retransported loess having ages from 3 Ma to about 130 ka and represents several periods of loess deposition and erosion. As many as 34 different tephra beds have been identified from Gold Hill Loess (Westgate et. al., 1990; Péwé, in press; S. J. Preece et. al., 1997, unpub. data) with dates from 2.01 Ma to 140 ka. Overlying the Gold Hill Loess and underlying the Goldstream Formation is the Eva Forest Bed of Sangamon age described in this report.

The Goldstream Formation overlies the Eva Forest Bed (Péwé, 1975b). It is a widespread deposit of perennially frozen, poorly bedded, organic-rich, gray-to-black retransported loess that is 10 to 35 m thick. The silt fills the valley bottoms and forms low-angle silt fans extending from valley bottoms onto the lower slopes (Fig. 4). Remains of mammoth, horse, and bison are common, but frozen carcasses are rare (Péwé, 1957; Guthrie, 1990). Palynological analyses and studies of mammal fossils indicate that trees were absent on the landscape during deposition of the Goldstream Formation (Matthews, 1968; Guthrie, 1968a, 1968b). All of these

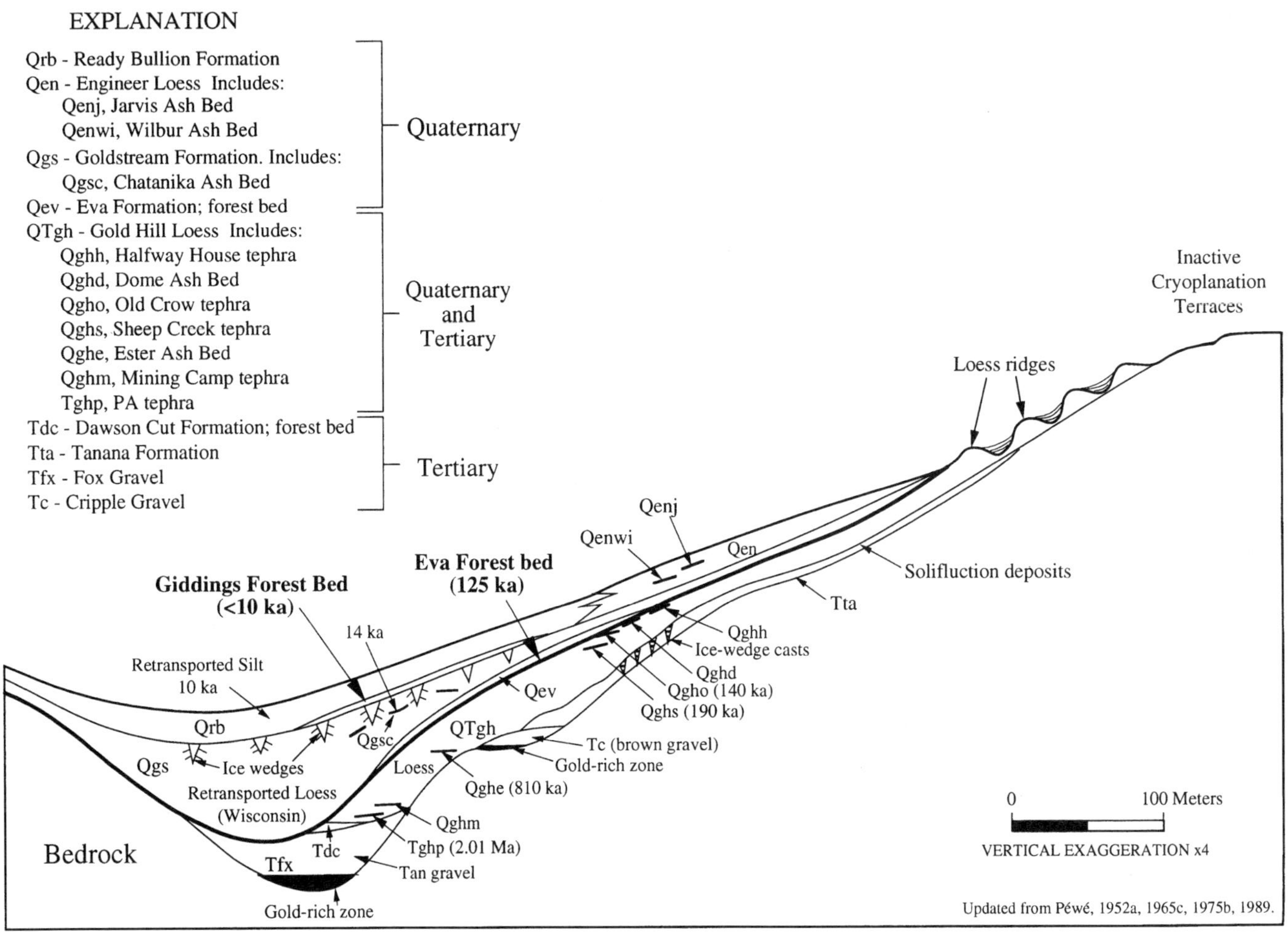

Figure 4. Schematic composite cross section of a creek valley, east central Alaska, illustrating stratigraphic relations of late Cenozoic deposits. Tephra in Gold Hill Loess identified and dated by J. A. Westgate, B. Stemper, and S. Preece. Halfway House tephra and Dome Ash Bed are both younger than Old Crow tephra, but which is the younger of the two is unknown. Dating of Sheep Creek tephra by Berger with TL method on loess. The unconformity between the Gold Hill Loess and overlying beds is emphasized by a heavier line.

characteristics, plus the presence of large ice wedges, are interpreted as indicating a harsh, arid, periglacial climate. The Goldstream Formation is Wisconsin in age (Péwé, 1965c, 1975b).

Unconformably overlying the Goldstream Formation is the Engineer Loess, located on the hillslopes, and the Ready Bullion Formation, which occupies the valley bottoms (Fig. 4). Both units are of Holocene age and have basal radiocarbon ages of about 10 ka (Péwé, 1975b). The Ready Bullion Formation is a poorly to well-stratified, perennially frozen, organic rich, retransported loess; it contains the Giddings Forest Bed, remains of a boreal forest less than 10 ka old (Péwé et al., 1995).

Modern climate

To understand more fully the environment and conditions at the time of the Eva Interglaciation in east-central Alaska, it would be well to review the climate, permafrost characteristics, and vegetation of the present Holocene Interglaciation. Fairbanks has the longest climatic weather record in central Alaska. A U.S. Weather Bureau station was established at the U.S. Department of Agriculture experiment station 5 km west of Fairbanks in 1904, and its records represent climatic conditions on the south-facing permafrost-free slopes at a ground elevation of 152.4 m. Since July 15, 1929, when the Weather Bureau Station was moved to Fairbanks, the records represent conditions on the flood plains of the Tanana and Chena Rivers at a ground elevation of 134.1 m (1929–1942) and 132.8 m (1942–present) above sea level. Records from 1930 to 1989 were used to compile Figure 5 (NOAA, 1989).

The Fairbanks region of interior Alaska has a distinctly continental climate, with a large variation of temperature from winter to summer (Fig. 5). Winters are normally long, dark, cold, and dry, and the temperature frequently drops to –48 to –45 °C. The lowest official air temperature of –52.2 °C was recorded in

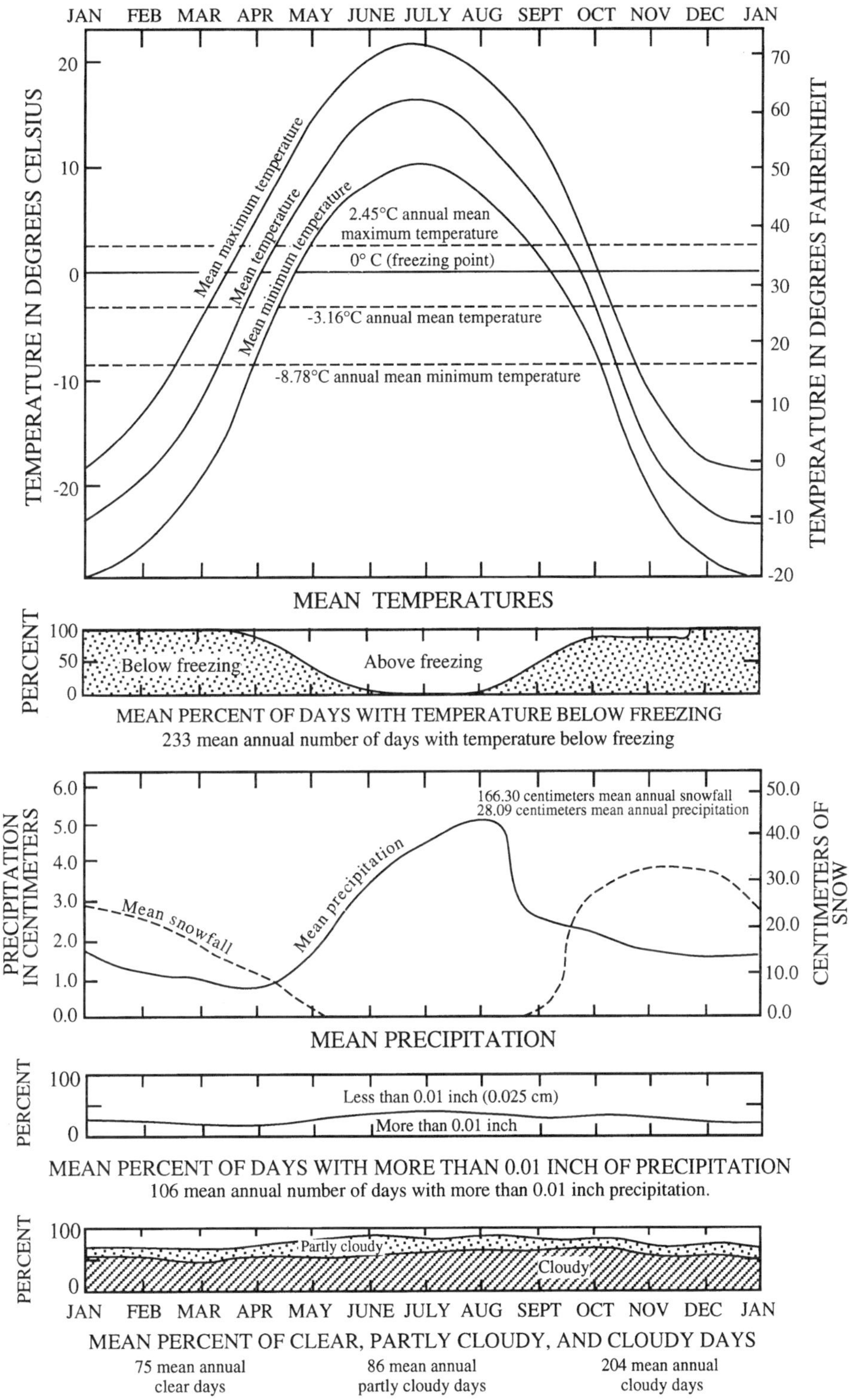

Figure 5. Climatic data for Fairbanks, Alaska, 1930–1989 (NOAA, 1989). Snowfall data, 1952–1989. Mean percent of days with temperature below freezing (U.S. Weather Bureau, 1943, p. 84).

December 1961. Summers are short, sunny, and warm, and maximum official air temperature was 35.6 °C in June 1969. The official mean annual air temperature on the floodplain in an urban environment (Fairbanks) is –3.2 °C (NOAA, 1989). The mean annual air temperature in low areas and the north side of hills is colder. For example, on the north side of College Hill (University of Alaska, Fairbanks) (Fig. 3) near Smith Lake, the mean annual air temperature is –6 °C or colder (Thomas Osterkamp, University of Alaska, June 6, 1990, oral communication). The mean number of days in Fairbanks with freezing temperature is 233, and freezing temperatures have been reported every month except July (U.S. Weather Bureau, 1943).

The wind regime in central Alaska is generally composed of a long, relatively calm winter period from September to May and a short, slightly windy summer period from June to August. A 10-year record at Fairbanks (U.S. Weather Bureau, 1943) indicates that the prevailing surface-wind direction for winter is north to northeast, and the prevailing summer direction is south to southwest. The average wind velocity in winter is about 1.6 to 1.8 m/s, and in summer is about 2.7 m/s. The annual mean wind velocity at Fairbanks is 2.19 m/s. High winds occur but are uncommon; about one gale is recorded yearly. The percentage of calms for the winter period is 6.3 and for the summer period, 3.3.

Mean annual precipitation for Fairbanks is 280 mm (NOAA, 1989). Precipitation normally reaches a minimum in spring and a maximum in August, when rainfall is common. Thunderstorms occur in Fairbanks on an average of about 8 days a summer, but most precipitation during the growing season falls as light showers. Sixty-three percent of the annual precipitation is from May through September. About 30% of the annual precipitation falls as snow, which normally covers the ground from middle to late October through mid April. The mean annual snowfall (1952–1989) is 166 cm. The maximum recorded snowfall of 374 cm occurred in the winter of 1990–1991 (Gunter Weller, University of Alaska, Fairbanks, May 3, 1991, written communication).

Present permafrost

The Yukon-Tanana Upland is in the discontinuous-permafrost zone of Alaska, and perennially frozen ground is widespread, except beneath hilltops and moderate to steep, south-facing slopes. Although some permafrost in the area is probably relict from colder Wisconsin-age conditions, perennially frozen ground forms today under favorable conditions. Sediments beneath the flood plain in the Fairbanks area are mainly sand and gravel and perennially frozen as deep as 80 m (Péwé, 1954). In this alluvium, permafrost frequently occurs as multiple layers of varying thickness, and in many areas permafrost is not present. Large ice masses are absent in the alluvium (Péwé, 1954, fig. 72, 1958, 1982; Péwé and Bell, 1974). In the retransported valley-bottom loess in creek valleys and on lower slopes permafrost is thickest at lower elevations—up to at least 110 m in the Yukon-Tanana Upland. It is in these slopes and creek-valley bottoms that remnants of the buried Eva Forest Bed are exposed. Permafrost in the retransported loess contains large masses of ice in the form of horizontal to vertical sheets, wedges, and saucer- and irregular-shaped masses. The ice masses are foliated (ice wedges) and range from less than 30 cm to more than 15 m in length (Fig. 6). Most of the ice is arranged in a polygonal or honeycomb network that encloses silt polygons 3 to 12 m in diameter.

About all ice wedges in central Alaska are now inactive (Péwé, 1966), except in local areas of particularly cold microclimates (Péwé, 1962; Hamilton et al., 1983). Active wedges exist in a cold microclimate in a peat bog at the base of a slope 1 km east of Farmers Loop Road and Ballaine Lake just north of the University of Alaska Fairbanks (Hamilton et al., 1983). The lake is a low spot in the terrain and in the winter the air temperature is 5° or 10 °C colder than terrain from about ⅓ km on either side of the lake (Thomas Osterkamp, April 21, 1992, oral communication). In the 1980s Thomas Osterkamp (unpub. data) recorded air temperatures at ground level and also frozen ground temperatures in the peat bog in the lowland 1 km south of the trenches described by Hamilton et al. (1983). Osterkamp estimates that the mean annual air temperature at this locality is about –6° (January 22, 1991, oral communication). Such temperatures should permit intermittent local ice-wedge growth. The temperature of permafrost in the Fairbanks area at the level of zero amplitude (depth of 8 to 15 m) is between about –0.5 °C to 0 °C (Péwé and Paige, 1963), although such temperatures as cold as –3.5 °C have been measured in local areas with especially cold microclimates (Thomas Osterkamp, February 22, 1991, oral communication).

Thawing of permafrost in flood-plain alluvium with low ice content results in little or no subsidence of the ground, but in creek-valley bottoms and on lower slopes, thawing of ice-rich retransported loess results in considerable differential subsidence of the ground surface and produces thermokarst topography. Many landforms indicative of permafrost with large ice masses are present in the vicinity of Fairbanks, including open-system pingos, low-center polygons, high-center polygons, thermokarst pits, beaded drainage, and thaw ponds and lakes (cave-in lakes; Péwé, 1982).

Modern vegetation

The Yukon-Tanana Upland is in the taiga, or northern boreal forest, of central Alaska (Fig. 7). The northern boreal forest consists of primarily open, slow-growing spruce interspersed with occasional dense, well-developed forest stands and treeless bogs (Viereck, 1973, p. 466). Viereck and Little state (1972, p. 9) that the extensive boreal forest of interior Alaska is composed of only three coniferous tree species—white spruce (*Picea glauca*), black spruce (*P. mariana*), and tamarack (*Larix laricina*); three large deciduous species—balsam poplar (*Populus balsaminifera*), quaking aspen (*P. tremuloides*), and paper birch (*Betula papyrifera*); and several species of willow (*Salix*); and two of alder (*Alnus*).

Vegetation in the lower parts of the Yukon-Tanana Upland is a complex mosaic resulting from a long history of repeated or periodic forest fires, from differences in slope exposure and parent

Figure 6. Typical large mass of foliated ground ice (ice wedge) exposed in retransported loess of Goldstream Formation during placer gold mining operations in Yukon-Tanana Upland, Dome Creek, 25 km north of Fairbanks, Alaska (photograph no. 1038 by T. L. Péwé, August 16, 1954).

material, and from a complicated pattern of permafrost (Viereck, 1975). On well-drained upland soils, where permafrost is lacking or is deeper than 1.2 m, large areas are covered by relatively young stands of paper birch and quaking aspen that have developed directly after forest fires or cutting. The successional aspen and birch stands will eventually be replaced by slowly growing white spruce now forming a scattered understory. The white spruce and the white spruce–paper birch forests are widespread on well-drained upland soils where there have been no wildfires in the past 200 yr. Upland areas underlain by shallow (0.3 to 1 m depth) permafrost are usually vegetated by black spruce in either open or dense stands. In these areas, black spruce often seed after fire, but may be preceded by stands of alder or paper birch.

Because the relationship between vegetation and permafrost is fairly clear in the Fairbanks area, vegetation can be used as a general indicator of permafrost conditions. Black spruce, larch, and bogs nearly always indicate the presence of permafrost within 0.6 to 1 m of the surface. White spruce and aspen usually indicate an area free of permafrost or an area in which the active layer is a meter or more thick. Paper birch occurs on sites that are free of permafrost or where the active layer has temporarily deepened as a result of burning or clearing.

Treeline on south-facing slopes is about 800 m elevation and is slightly lower on north-facing slopes. Just below altitudinal treeline, white and black spruce typically form an open woodland. Above 1,000 m elevation in the southern Yukon-Tanana Upland, hill slopes and summits and ridge flanks and crests bear a discontinuous carpet of low-growing herbs, grasses, sub-shrubs, cushion plants, and lichens. This cover is interrupted by rocky rubble and contains low, dense, pruned thickets of willows and resin birch in hollows where snow collects each winter.

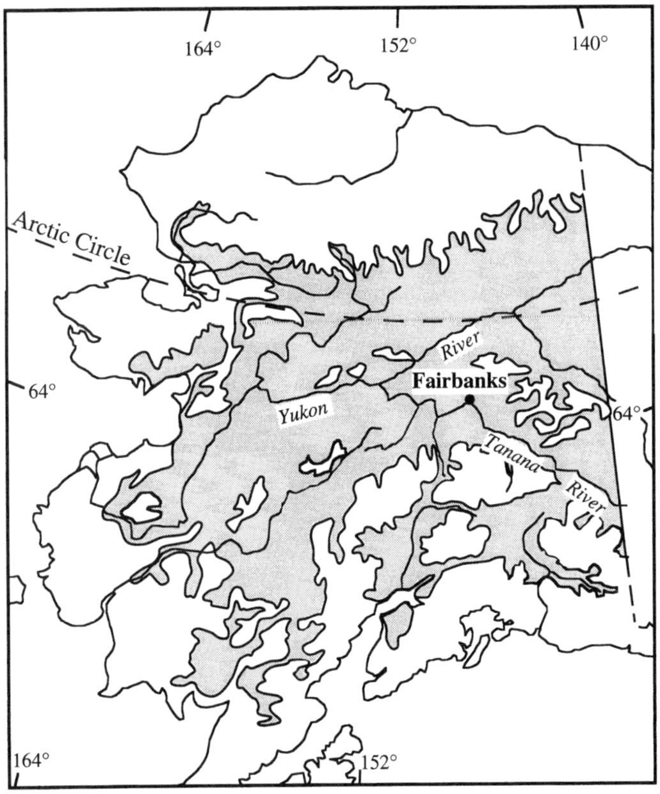

Figure 7. Shaded area shows extent of present-day boreal (taiga) forest in Alaska (from Viereck and Little, 1972).

EVA FOREST BED

Description

The Eva Forest Bed represents an ancient boreal forest and has been identified at many localities in the Yukon-Tanana Upland (Figs. 1–3). Péwé has seen all reported exposures except that at Birch Creek on the northern edge of the upland (Edwards and McDowell, 1989, 1991) and the Palisades exposure on the Yukon River a few miles west of the upland (Fig. 2; Begét et al., 1991). Only macrospecimens of spruce and birch were collected by Péwé during the period from 1949 to 1993 at the visited localities. Locally *Sphagnum* is present, as well as unidentified small stems, probably willow, cottonwood, and aspen.

Eva Forest Bed as a stratigraphic unit. The Eva Forest Bed[2] is a 0.5- to 1-m-thick bed of peat lenses, sticks, logs, rooted and unrooted stumps of trees in retransported loess that is rich in minute carbonized plant fragments. Locally, the perennially frozen soil, silt, and organic clasts are tightly matted into a relatively resistant layer. For example, at Dawson Cut the 15- to 30-cm-thick forest bed is a tightly interlocked mass of prostrate limbs and logs of white spruce and paper birch, both with bark, and white spruce cones. Here the bed forms a resistant layer 3 m below the modern ground surface. Here also the forest bed unconformably overlies a tilted, green Gold Hill Loess and a thin unidentified tephra layer. Clear ground ice has formed at the loess-wood contact and between wood layers (Table 1, Fig. 8). In practically all of the fresh mining-cut exposures in the Fairbanks area, this resistant layer of the Eva Forest Bed constitutes a poorly to well-developed topographic bench between the underlying Gold Hill Loess and the overlying Goldstream Formation. Good examples of the bench structure occur at Eva Creek, Dawson Cut, and Gold Hill mining cuts (Figs. 9, 10; see also Matthews, 1968, fig. 2, 1970, fig. 5; Guthrie, 1968a, fig. 2; Péwé, 1952a, fig. 37, 1975a, fig. 30, 1975b, fig. 6).

The appearance and preservation of the Eva Forest Bed varies from exposure to exposure. Locally, it is present as an extensive buried forest bed hundreds of meters long, especially on top of low hills (Gold Hill; Table 1) and on the higher parts of lower slopes, or, only as local patches. Elsewhere, the bed may exist as (1) isolated logs and stumps grading into just a single stump, (2) simply wood fragments (Figs. 11, 12), or (3) a woody silt as at Sheep Creek (Fig. 13).

The silt in the Eva Forest Bed is a blackish retransported loess (Fig. 14c). For the most part, the silt associated with the logs and stumps is indistinguishable from the retransported loess of the overlying Goldstream Formation. Mechanical analyses of the silt show a slightly higher percentage of clay-size particles in the Forest Bed–Goldstream loess than the underlying Gold Hill Loess. This is illustrated by grain-size analyses of the silts in Eva Cut (Figs. 14c, 15). These two analyses mirror many cumulative-frequency curves of grain-size analyses of silt near the contact of Goldstream loess and Gold Hill Loess from many exposures in the Yukon-Tanana Upland (Péwé, 1955; field notes).

Under ideal undisturbed conditions, the "soil" of the Eva Forest Bed would be formed from the underlying Gold Hill Loess. Under conditions of some erosion and then complete burial by retransported, dark Goldstream Loess, and cryoturbation by seasonal frost action, the silt deposited around the logs and stumps could represent a later time of a colder and drier environment. At Eva Creek, a tundra environment is indicated by (1) fossil insects (Matthews, 1968), (2) pollen (Matthews, 1970; Hopkins, 1982), and (3) remains of small mammals (Guthrie, 1968b) from the silt (loess) associated with the white spruce macrofossils of the Eva Forest Bed.

Trees of the Eva Forest Bed.

Species. Spruce dominated the tree remains at almost all of the exposures of the Eva Forest Bed. The largest spruce samples were as much as 32 cm in diameter and 3 m long (Fig. 16). The abundance and size of spruce remains appears to be consistent with the modern boreal forest, where spruce is the most common and the largest tree present (Viereck, 1973; Lutz, 1956, p. 8–9). Most of the spruce remains are white spruce (*Picea glauca*) rather than black spruce (*Picea mariana*) as indicated by (1) the presence of white spruce cones and nee-

[2]The terms Eva Forest Bed and Eva Formation have been used concurrently since the 1940s. Later, the term Eva Formation was formalized (Péwé, 1975b, p. 14), with the type locality being at Eva Creek (Fig. 3). In this report the term Eva Forest Bed is used.

TABLE 1. BASIC DATA FOR LOCALITIES AND WOOD SAMPLES OF EVA FOREST BED NEAR FAIRBANKS, ALASKA

Sample Number	Location	Date	Specimen	Description	Tree Rings	Age ^{14}C	Identified	Discussion and Stratigraphy*	Comments
None	Gold Hill	8/20/48	Stump	20 cm diameter, leaning in place, some burnt wood			Spruce (Péwé)	Forest bed crops out continuously for 3/4 km in upper part of Gold Hill Cut, 7 m below surface; base of Goldstream Formation; overlies Gold Hill Loess	
None	Eva Creek	6/28/49						First description of unconformity at top of Gold Hill Loess and the overlying Goldstream Formation	Figures 37, 38, Péwé, 1952a
None	North side of Ester Island	7/12/49	Forest bed	Many sticks of burnt birch			Birch (Péwé)	Overlies unconformity with green Gold Hill Loess	Figure 39, Péwé, 1952a. Figure 22, this report
None	Dawson Cut	8/7/49	Logs	Logs and 7.5 cm-diameter sticks				Logs at base of Goldstream Formation unconformably overlying green silt and brownish tephra in Gold Hill Loess	
None	Dawson Cut	9/9/49	Forest bed	Interlocked tight mass of prostrate limbs and logs of white spruce with bark and white spruce cones			Spruce and birch with bark (Péwé)	Forest bed 15 to 30 cm thick forms a resistant mass 3 m below surface and lying unconformably on green Gold Hill Loess and tephra. Clear ground ice has formed at contact and in wood layers	Good photos. High up slope, not in valley bottom. Figure 8, this report
None	Eva Bench Cut (150 m west of Eva Creek)	9/24/49						Excellent photo panorama of east wall of mining exposure showing contact between Goldstream Formation and Gold Hill Loess (green) and location of Eva Formation	Good photos. Figure 35, Péwé, 1952a. Figure 21, this report
3	Amy Creek	6/3/51	Stump	"Large" tree stump. Surface of wood is charred black	71		Spruce (UAZTRL, Leavitt)	Stump in place at base of Goldstream Formation on underlying gravel 8 m below top of Goldstream Formation and 10 m below surface	Figure 23, this report

TABLE 1. BASIC DATA FOR LOCALITIES AND WOOD SAMPLES OF EVA FOREST BED NEAR FAIRBANKS, ALASKA (continued, p. 2)

Sample Number	Location	Date	Specimen	Description	Tree Rings	Age ^{14}C	Identified	Discussion and Stratigraphy*	Comments
9	Dawson Cut	6/9/51	Rooted stump	Well-developed forest bed over a "considerable" distance; stumps in place. Charred wood	90	>29,000 yr (L137P)	Spruce (UAZTRL)	Well-developed forest bed under Goldstream Formation and unconformably over green Gold Hill Loess and 15 cm-thick white, glassy tephra, 11 m below surface	Overlies about 31 m of Gold Hill Loess. See Figure 10, this report
19	Lower Eva Creek-east wall	6/4/52	Log fragments	Many fragments 30 cm long, 1 cm thick and 2 cm wide from fracturing of large spruce log. Ring widths 1 to 1.5 mm. Fragments from surface black and burnt. Inside fragments not burned	19+	>30,000 yr (L163J)	Spruce (UAZTRL)	Concentration of wood fragments in Eva Formation, overlying green silt of Gold Hill Loess	Figure 46, Péwé, 1952a. Figure 11C, this report
20	Dome Creek	6/8/52	Rooted stump	Burned stump. Charred bit of bark as well as surface and wood under bark	24+		Spruce (UAZTRL)	Stumps in place at contact with underlying Gold Hill Loess; 8 m below surface	Upper end of Dome Creek Cut. Figure 9, this report
21	Dome Creek	6/8/52	Log	Small log and fine schist fragments. Few wood fragments have surface burnt	32		Birch (UAZTRL)	Log and schist fragments on top of unconformity overlying green Gold Hill Loess 7 m below surface. Thin ash layer under wood	Figure 9, this report
95	Middle Sheep Creek Cut	9/20/55	Woody silt	Small wood fragments				Silt with abundant wood fragments lying on unconformity above Gold Hill Loess	Figure 13, this report
178	Eva Creek	7/2/67	Rooted upright stump	Stump from forest bed in black silt			Not recorded, probably spruce	Stumps in forest bed at unconformity at top of green Gold Hill Loess and tilted tephra layer. Section sampled for geochemical analysis of silt	Péwé 1975a, p. 59-60. Sample lost
None	Upper Eva Creek	7/8/70	Log	Log				Log at unconformity with underlying Gold Hill Loess	No specimens collected. Section 40 m high. Silt sampled for geochemical analysis (See Fig. 6, Péwé 1975b; Fig. 29, Péwé 1975a. See p. 59–60, Péwé, 1975a)

TABLE 1. BASIC DATA FOR LOCALITIES AND WOOD SAMPLES OF EVA FOREST BED NEAR FAIRBANKS, ALASKA (continued, p. 3)

Sample Number	Location	Date	Specimen	Description	Tree Rings	Age C^{14}	Identified	Discussion and Stratigraphy*	Comments
186	Upper Eva Creek west wall	6/29/87	2 cm-diameter spruce stick. 10 cm-diameter birch	Spruce sticks and birch 10 cm-diameter log with bark; burnt wood also under no. 187	Too broken		Spruce and birch (Péwé, U.S. Forest Service, UAZTRL)	Wood fragments near and under large log No. 187	Figure 14C, this report
187	Upper Eva Creek west wall	6/29/87	Prostrate log	31.8 cm-diameter prostrate log sloping toward valley bottom; large, 1 mm wide rings; excellently preserved	144	>34,335 yr >41,200 yr (Beta 33074) >70,000 yr (A. Long)†	White spruce (Péwé, UAZTRL, U.S. Forest Service)	1 m of log sticking out of valley mining cut. 1 m sawed off. "Long" log still in ground. Lies in Eva Forest Bed near unconformity with underlying Gold Hill Loess	Wood records evidence of bark beetles (Scolytidae). Common today on white spruce in central Alaska. (See Long and Kalin, 1992a, 1992b; McCormac et al., 1993 for C^{14} dating) cover photograph, this report
188	Upper Eva Creek west wall	6/29/87	15 cm-diameter log	15 cm-log near 15 cm diameter upright stump	Too broken		Spruce (UAZTRL)	Log and stump in retransported brownish silt of Eva Formation; sharp contrast with underlying massive, inorganic Gold Hill Loess and Old Crow tephra	Figure 14C, this report
193	Upper Eva Creek west wall	5/6/91	10 cm-diameter birch log	Oval birch log; bark well preserved	15+1		Birch (Péwé, UAZTRL)	At right angle to and directly under large log no. 187	
198	Upper Eva Creek west wall	6/20/91	16 cm-diameter log	26 cm-long; part of large 35 cm stump with roots. Outside of log charred	179	>42,410 yr. (Beta 46130)	Spruce (Péwé, UAZTRL)	Next to no. 188. Stump in growth position on unconformity	No bark but surface highly riddled by bark beetle channels (Spruce beetle pupal chambers; Werner et al., 1977, Fig. 5) Figures 18, 19, this report

*All localities have stratigraphic diagrams in field notebooks.
†Austin Long, U. of Arizona, oral communication, January 4, 1993.
UAZTRL = University of Arizona Laboratory for Tree Ring Research, Tucson, AZ 85721. Peter Brown identified specimens and determined number of tree rings. All samples collected by T. L. Péwé.

Figure 8. Perennially frozen interlocked tight mass of prostrate limbs and logs of white spruce with bark and cones, and birch limbs of Eva Forest Bed unconformably overlying green Gold Hill Loess with tephra. Cylindrical piece of birch bark on right side extracted from mass and exhibited for photograph. Dawson Cut, 13 km north of Fairbanks, Alaska (photograph 457 by T. L. Péwé, September 9, 1949).

dles, (2) the larger size of the stem remains, and possibly, (3) evidence of spruce bark beetles.

Remains of birch trees are common in the Eva Forest Bed. Logs as much as 10 cm in diameter and as much as 30 cm long are present. The well-preserved bark suggests that the most common species is thought to be *Betula papyrifera* (Fig. 8). Specimens of willow and poplar are small and not very abundant. Perhaps the only exception to this is the presence of willow and poplar with spruce reported by Edwards and McDowell (1989, 1991) from a 45 m high exposure of loess and retransported loess along Birch Creek, near Circle, on the northern edge of the Yukon-Tanana Upland at the southern edge of the Yukon Flats (Fig. 2).

The Birch Creek locality is different from other exposures of Eva Forest Bed in the Yukon-Tanana Upland: (1) This exposure is in the southern marginal upland of the Yukon Flats adjacent to the Yukon-Tanana Upland along a natural exposure of Birch Creek. All other exposures are in mining cuts in valley bottoms of small streams draining the upland. (2) This deposit of buried trees of a boreal forest exposed at Birch Creek is a collection of tree remains that tumbled into a thermokarst lake, perhaps during a climatic warming. (3) The forest bed here contains the largest preserved remains of willow and poplar and exhibit well-preserved beaver-chewed ends. The logs are older than 58,000 years and lie about 5 m stratigraphically above the Old Crow tephra (Edwards and McDowell, 1989, 1991).

Counts of tree rings on spruce from the Eva Forest Bed indicate that the trees were from 10 to 179 years old when they died (Table 1, cover photograph). The rings are excellently preserved

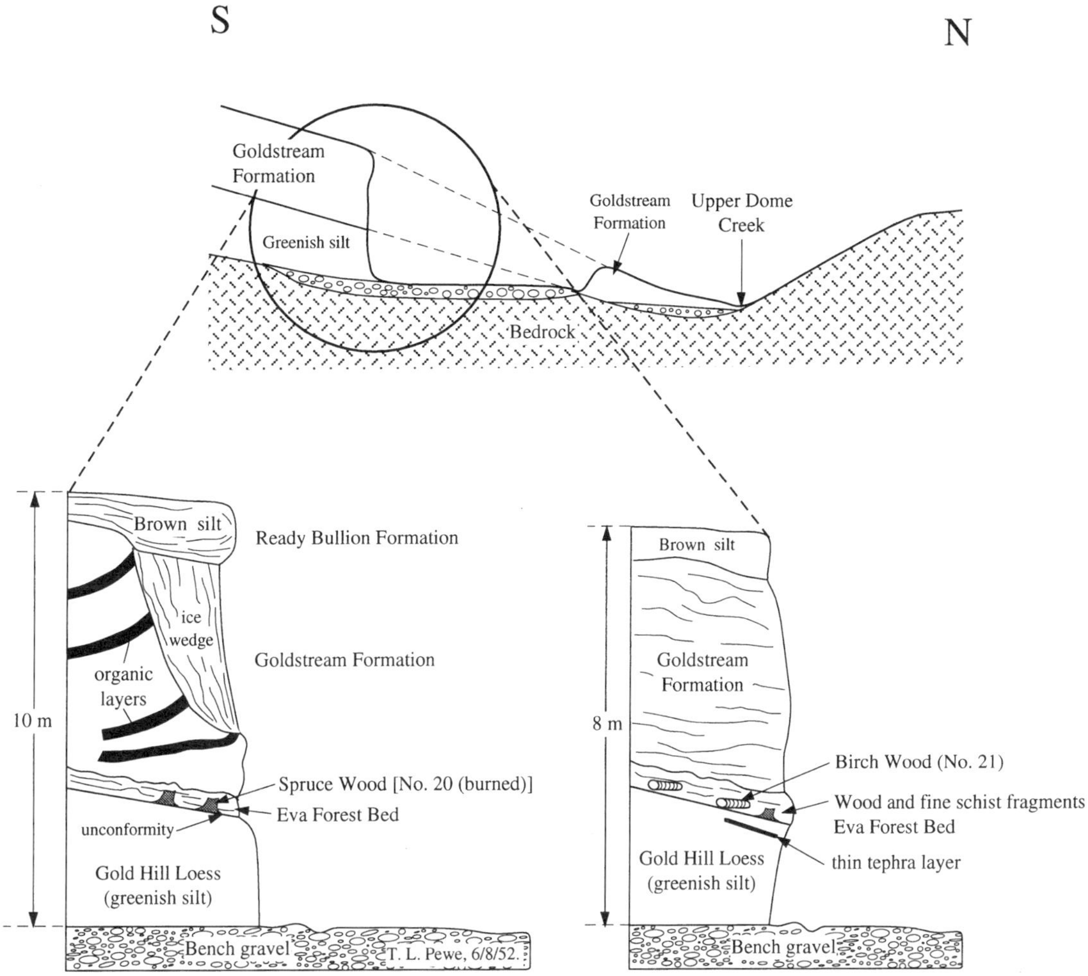

Figure 9. Stratigraphy of two exposures of perennially frozen Quaternary sediments on south wall of mining cut on upper Dome Creek, 25 km north of Fairbanks, Alaska. Eva Forest Bed with logs and stumps of spruce and birch unconformably overlies greenish Gold Hill Loess.

in many specimens and average 1 to 1.5 mm thick. A cut and polished stem disk of log no. 187 from Eva Creek (Fig. 14a, cover photograph) illustrates rings very similar to rings of a stem disk from a section of modern white spruce (no. M-2) cut near Fairbanks (Fig. 17; Table 2) from about the same elevation. The Eva Forest Bed specimen has 144 rings, and the modern spruce specimen has 113. A cut and polished stem disk of spruce log no. 198 from Eva Creek (Figs. 14c, 18) has 179 rings, most of which are extremely narrow. Scientists at the U.S. Department of Agriculture Forest Service, Forest Products Laboratory, in Madison, Wisconsin, who have studied the specimen state that it is not possible to differentiate between white and black spruce from wood anatomy.

Method and degree of preservation of wood specimens. Logs, stumps, roots, cones and a host of wood fragments lying on the floor of the Eva Forest were buried by silt. On slopes, some wood remains were probably buried rapidly as layers of silt from the Gold Hill Loess or early Wisconsin loess were washed down hill. At other sites, wood remains were gradually buried by early Wisconsin airfall loess accumulating over macrospecimens. Some weathering and decay of the buried specimens probably continued until the silty organic bed became perennially frozen, and likely remained frozen until exposed by mining operations.

Spruce wood is the best preserved of the tree species in the buried forest; when some of the logs are thawed and dried, they strongly resemble modern wood (Fig. 16). A significant fraction of the spruce wood is hard and cohesive, but most is cracked and tends to split upon drying. The bark is not commonly preserved, but some white spruce cones are well preserved. The wood of the birch tree is generally not as well preserved as the spruce, but the white birch bark is locally very well preserved (Fig. 8). None of the wood is mineralized, but is stained gray to light tan. Many fragments are black from burning. Wood tentatively identified as willow or poplar is not widespread. It consists of small specimens with no bark and is soft when wet and splits upon drying.

How does the preservation of the wood in the Eva Forest Bed compare with the preservation of the wood in the other two regional forest beds present in the frozen loess sections of the

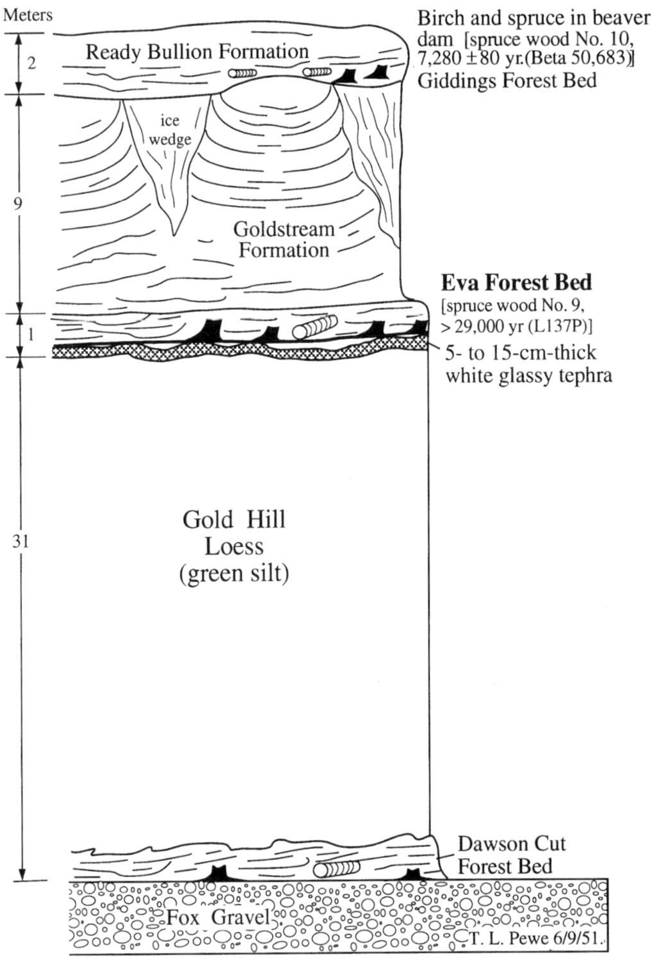

Figure 10. Sketch of stratigraphy of perennially frozen late Cenozoic loess and forest beds at Dawson Cut mining exposure, 13 km north of Fairbanks, Alaska. The 5- to 15-cm-thick white glassy tephra immediately under the Eva Forest Bed is possibly the Old Crow tephra because of its thickness.

Yukon-Tanana Upland: the older (2 Ma) Dawson Cut Forest Bed and the younger (1–10 ka) Giddings Forest Bed (Péwé et al., 1995; Figs. 10, 11)? The wood of the frozen Eva Forest Bed is moderately to excellently preserved. It is not compressed, smashed, or flattened, and wood cell structure is good. Such preservation is in great contrast to spruce wood of the underlying Dawson Cut Interglaciation Forest Bed (Péwé, 1952a, p. 106–111, 1975b, p. 8–9; Péwé et al., 1995). This older wood is more weathered, smashed, flattened, and iron-stained than specimens in the Eva Forest Bed. The cell structure is not well preserved. Birch and poplar are rare. The frozen tree remains of the Giddings Forest Bed (Péwé et al., 1995) on the other hand, are all excellently preserved and therefore also differ from the Eva Forest Bed. As Giddings (1938, p. 4) stated, the wood differs from modern wood only in color and some of the frozen leaves are still green (Péwé, 1975b, p. 20). The radiocarbon dates of wood from this forest bed, including those of wood collected by Giddings from the type locality, are from 10,000 to 1,000 years old (Giddings, 1938; T. L. Péwé, unpub. data; Péwé et al., 1995; Table 2). Many of the specimens (33%) used in the comparative study of $^{13}C/^{12}C$ isotope ratios of Eva Forest Bed trees with buried Holocene and modern trees presented later in this report, are from the Giddings Forest Bed. Therefore, it probably would be well to formalize the term "Giddings Forest Bed" (T. L. Péwé, field notes; Péwé et al., 1995). This Holocene forest bed is widespread in the Yukon-Tanana Upland (Figs. 10, 11) and is here named after the late J. Louis Giddings, pioneer archaeologist and dendrochronologist who worked in Alaska and first described this wood. The type locality of this forest bed is in the Ready Bullion Formation (Fig. 7) at the mouth of Engineering Creek (Fig. 3). There is a far greater number of preserved willow and poplar in the Holocene forest than in the two older forests, but this may be in part an artifact of selectively collecting many beaver-chewed specimens from the Holocene forest bed (Table 3, Fig. 10) for study.

The downslope and valley-bottom axis orientation of many of the logs of the Eva Forest Bed indicates some retransportation of the logs downslope after they broke off from the stumps. Some of the rooted stumps are still upright (Table 1), others are "leaning," and still others are no longer rooted but have fallen over or tumbled down hill.

Absence of beaver-chewed wood. No beaver-chewed wood is reported from the Eva Forest Bed in the many gold mining exposures in valley bottoms of small streams draining the Yukon-Tanana Upland. The absence of such wood is in great contrast to locally abundant beaver-chewed wood scattered on the floors of some mining sites and also exposed in old buried beaver dams fairly common in the perennially frozen silt walls of some of the cuts. Historically miners, scientists, and tourists have collected hundreds of these well-preserved fossil, beaver-chewed wood specimens over the years that placer gold mining was active in the Yukon-Tanana Upland (Giddings, 1938; Péwé, 1952a, fig. 48; Table 3).

We present here for the first time the concept that it appears that all the beaver dams and beaver-chewed wood cited over the years from the mining cuts are from the Giddings Forest Bed. The wood is all Holocene in age and from the Reading Bullion Formation (Fig. 4). Many radiocarbon dates are now available on such wood, and all known dates are 10,000 years or younger (Péwé, 1975b; Table 4; Table 3; Figs. 10, 12).

Although spruce and birch do occur in beaver dams, most wood is willow or poplar. This is because such trees were common along small streams then as they are today (Table 3), and also, beaver favor these trees for their diet (Murray, 1961; Hakala 1952; Robert Hufman, professional beaver trapper, Fairbanks, Alaska, August 21, 1995, oral communication).

As mentioned earlier, beaver-chewed wood does exist in at least one of the two Eva Forest Bed localities adjacent to the Yukon-Tanana Upland. At the Birch Creek locality on the north edge of the upland (Fig. 2) next to the Yukon Flats, Edwards and McDowell (1991) report beaver-chewed wood. Wood of the forest bed, exposed at the Palisades of the Yukon River, just west of the Yukon-Tanana Upland (Fig. 2), has not been examined but viewed only from a distance from the river

(Begét et al., 1991). It is not known if beaver-chewed wood exists there.

The absence of beaver dams and beaver-chewed wood in the Eva Interglaciation Forest Bed in mining cuts in the small stream valleys of the Yukon-Tanana Upland may indicate the following: (1) Perhaps during the last interglaciation beavers did not invade the small streams (creeks) and existed only along major and minor rivers (Tanana, Yukon, and others). (2) An alternative interpretation is that the forest bed was rarely preserved at the bottom of creek valleys because downcutting of the loess during interglacial time removed most of the pre-Wisconsin valley-bottom silt, as well as beaver dams, as the streams eroded to underlying gravel (Fig. 4). Or (3), field observations appear to indicate that wood from willow and poplar trees is much less common and less well preserved in the frozen silt of the older deposits than in the younger. Therefore, most beaver-chewed wood older than 100 ka may have weathered away.

Forest fires. In his thorough work on the ecological effects of forest fires in interior Alaska, Lutz (1956) emphasizes that the boreal forest is very susceptible to local destruction by fire. The occurrence of extensive and repeated fires in prehistoric, historic, and modern time is well substantiated (Lutz, 1956, p. 9; Shostakovitch, 1925; Viereck, 1973). Hundreds to thousands of km^2 are burned almost every summer, not only in Alaska, but also in Russia and Canada. The fires are caused by man and by lightning. Shostakovitch (1925) describes the fires of 1915 in Siberia that were estimated to have burned 88,000 km^2 of forest. Smoke from these fires is estimated to have spread over about 4,160,000 km^2, an area the size of the whole of Europe and more than four times the area of Alaska. Smoke from forest fires today commonly blankets the city of Fairbanks, and air traffic is occasionally diverted because of it.

We report here for the first time the probable presence of widespread forest fires in the Yukon-Tanana Upland during the time of the Eva Interglacial Forest. Although only a few burnt wood specimens are listed in Table 1, almost every one of the many exposures of the forest bed noted from 1948 to 1993 in the Yukon-Tanana Upland yielded evidence of former fires. This evidence can be grouped into two types. First is the presence of rooted burnt stumps, fallen logs, and sticks, mainly spruce and birch (Table 1). Fire charring is on bark, and on the debarked surface of the tree. The macro remains of forest fires were apparently buried by accumulation of silt, retransported loess, and then frozen. A second type of evidence is the ubiquitous accumulation of small carbon particles and charcoal fragments (1 to 10 mm in diameter) in the silty soil, probably produced as the burnt tree stems and logs are weathered prior to burial. Locally, the burnt debris forms 1- to 3-mm-thick layers in the soil of the Eva Forest Bed.

Bark beetles. We report here, also for the first time, evidence for the existence of spruce bark beetles (Scolytidae) during the last interglaciation in Alaska. Small, meandering beetle galleries about 3 to 4 mm wide on the debarked surface of white spruce logs of the Eva Forest Bed were first observed by Péwé and H. J. Lutz in the Fairbanks area in 1957 (Table 1, Fig. 19). Péwé has subsequently examined many beetle-scored specimens from different exposures. Werner et al., (1977, p. 3, 4; Figs. 4 and 5) reports the following for Scolytidae scoring on modern white spruce trees in central Alaska:

> The female beetle bores through the bark into the phloem area, i.e., the area between the outer bark and the wood, and starts to construct an egg gallery, scoring the xylem (wood) slightly in the process. . . . The egg gallery is constructed parallel to the wood grain. . . . The galleries are slightly wider than the beetle. Eggs are laid in groups on either side of the egg gallery. . . . Newly-hatched larvae bore at right angles to the longitudinal egg gallery and feed in clusters in the phloem.

Because of the excellent wood preservation of the interglacial white spruce logs, in some specimens the bark beetle galleries can hardly be distinguished from those in modern white spruce logs (Fig. 19).

Evidence for bark beetles in the Eva Forest Bed is only from the largest or oldest trees, for example samples 187 and 198 (Table 1). Bark beetles are the primary cause of white spruce mortality in central Alaska today (Bright, 1976; Werner et al., 1977; Werner and Helsten, 1984), and it is the large, mature white spruce stands that are highly susceptible to spruce beetle attack (Fig. 6; Baker and Kemperman, 1974).

Bark beetle fossils are known from deposits of late Wisconsin and Holocene age in North America from Colorado (Elias and Edwards, 1982), Ontario (Ashworth, 1977), and Wisconsin (Morgan and Morgan, 1979), and elsewhere. Matthews et al., (1990) reports at least four genera of bark beetles from the last interglacial deposits along the Porcupine River (Fig. 1) in northern Yukon Territory, Canada.

Pollen. Unfortunately little additional information about the Eva Forest Bed is yet available from a study of its pollen. This is perhaps due to the very few palynologists studying the Quaternary silt deposits of central Alaska during the 1940s and early 1960s when much of the field work for this report was underway. Also, pollen is reported to be sparse and poorly preserved in the loess and, for the most part, it is the loess sections that were heavily sampled.

Pollen investigations in central Alaska of the past two decades have dealt mainly with late Wisconsin and Holocene sediments and modern vegetation. However, there is one notable exception: the detailed work of Edwards and McDowell (1991) on silt deposits of Eva Forest Bed age at Birch Creek near Circle City near the Yukon River (Fig. 2). The pollen assemblages in that Eva Forest Bed indicate a closed boreal forest similar to a modern forest with *Picea, Betula, Alnus, Populus, and Salix*. Pollen of *Typha latifolia* and aquatic taxa such as *Myriophyllum* also occur. *Typha latifolia* is near its northern limit in the area today (Edwards and McDowell, 1991, p. 48). This common cattail grows near Fairbanks today.

Distribution

Localities exhibiting some aspect of the Eva Forest Bed are known throughout the Yukon-Tanana Upland (Fig. 2). Most

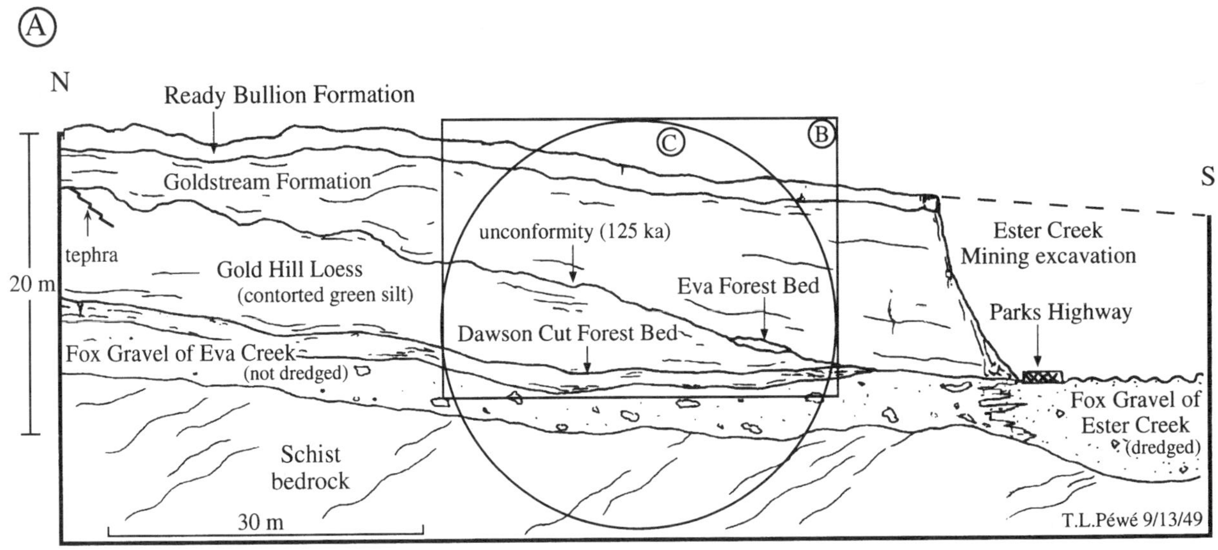

Figure 11 (on this and facing page). Late Cenozoic stratigraphy of perennially frozen sediments and forest beds exposed in the east wall of the lower Eva Creek mining cut, 16 km west of Fairbanks, Alaska. (A) Longitudinal sketch; (B) Area covered by photograph nos. 459–460 by T. L. Péwé, September 9, 1949 (Péwé, 1952a, fig, 46); (C) Details of the stratigraphy, with radiocarbon dates, September 8, 1949. See Figure 3 for geographical relationship between Ester Creek and its tributary Eva Creek.

exposures are in thick loess deposits in the many mining cuts near Fairbanks (Figs. 2, 3; Péwé et al., 1995). In addition to the several localities described earlier (Figs. 2, 3, Table 1), Péwé made reconnaissance observations of mining cuts in loess and retransported loess in the areas near the Alaska-Yukon Territory border (Lost Chicken Creek, e.g., Fig. 20), and in the Dawson Mining District in Yukon Territory where "vegetation" layers resembling the Eva Forest Bed were noted (T. L. Péwé, 1955, unpub. data). The loess stratigraphy there has similarities to that in the Fairbanks area. Also many small exposures of the Eva Forest Bed examined in the Fairbanks area over the years were not documented in detail, such as

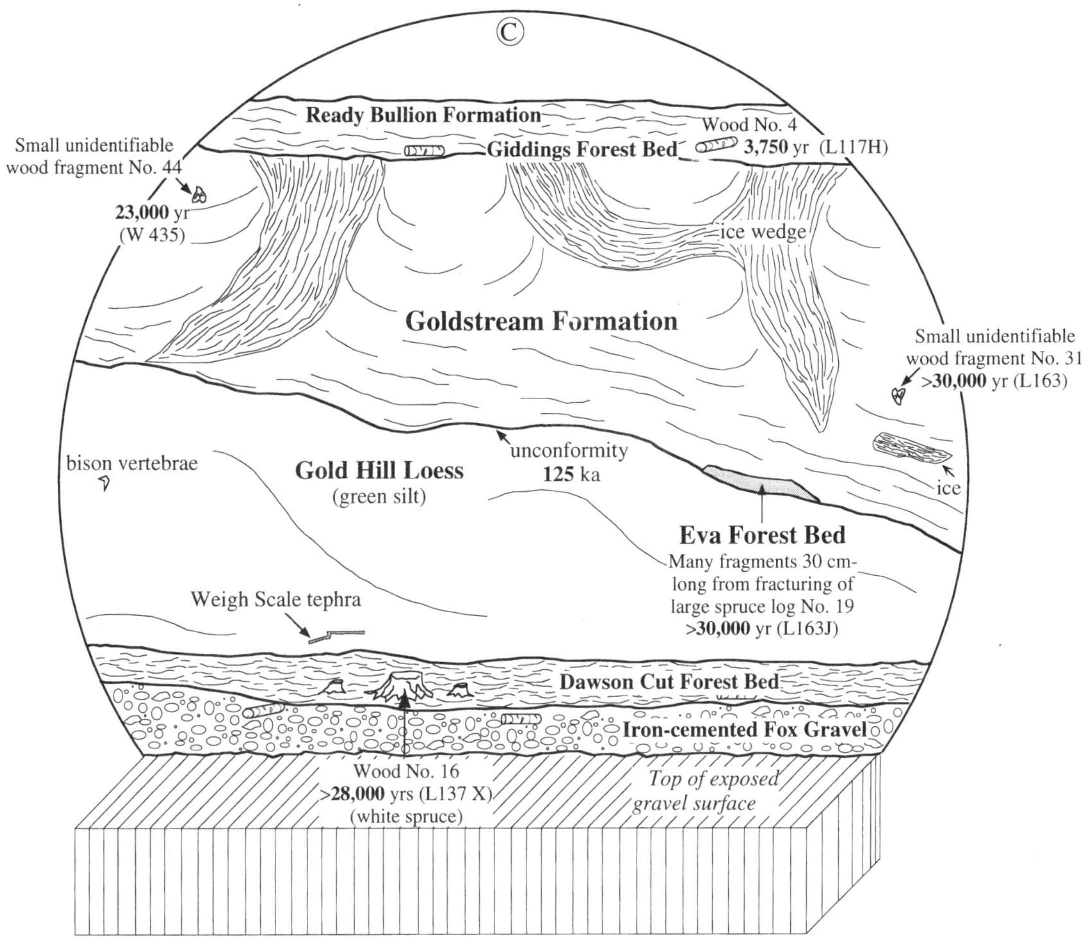

those at Fairbanks Creek, West Dawson Cut, Ready Bullion Creek, and others.

The westernmost reported exposure (Palisades site) of what appears to be the Eva Forest Bed is an organic rich zone with spruce logs as much as 20 cm in diameter and 3 m long in perennially frozen loess 4 or 5 m above the Old Crow tephra. This locale is at the Yukon River (Begét et al., 1991). This site is a few kilometers west of the Yukon-Tanana Upland on the south bank of the Yukon River near Tanana (Yeend, 1977), 80 km west of Manley Hot Springs (Fig. 2). Begét et al. (1991) tentatively correlated the buried wood horizon with the Eva Forest Bed. Overall, exposures of the buried Eva Forest Bed are limited to creek-valley mining exposures at elevations of 100 to 300 m and to exposures near the Yukon River (Figs. 2, 3). Therefore, no information is available on the upper limit of this fossil forest.

Stratigraphy

General statement. Hundreds of stratigraphic sections of frozen Quaternary sediments were seen over the past four decades when fresh gold mining exposures were available in the lower creek valleys of the Yukon-Tanana Upland (Fig. 2). The most conspicuous stratigraphic feature by far in almost all of the exposures is the prominant unconformity between the Goldstream Formation and/or Eva Forest Bed and the Gold Hill Loess (frontispiece, Fig. 21—see center foldout). There is usually a great contrast between the overlying blackish, ice wedge–rich retransported Goldstream Formation and underlying frozen green or brown massive Gold Hill Loess. Commonly, the Goldstream Formation with relatively poor bedding parallel to the surface is in sharp contrast to the tilted or folded loess and tephra below an angular unconformity (Figs. 4, 8, 9, 11, 12, 13, 14; also, see figures in Péwé, 1952a). As illustrated in Figures 4 and 11, the Goldstream Formation of Wisconsin age thickens downslope into creek-valley bottoms, whereas the underlying Gold Hill Loess of pre-Wisconsin age becomes thinner downslope because of erosion and is generally absent in part or all of the creek-valley bottoms. Stratigraphically, proceeding downhill to creek-valley bottoms, older parts of the Gold Hill Loess with enclosed tephra progressively underlie the Goldstream Formation. In most creek-valley bottoms, the Goldstream Formation lies directly on the gold-bearing Fox Gravel of late Tertiary age (Fig. 4). The stratigraphic relations outlined above indicate that a major period of erosion and climatic warming occurred at the end of deposition of Gold Hill Loess (Péwé, 1952a, 1957, 1965c, 1975b, in press). This event is summarized below.

With the retreat of glaciers in the high mountains, major

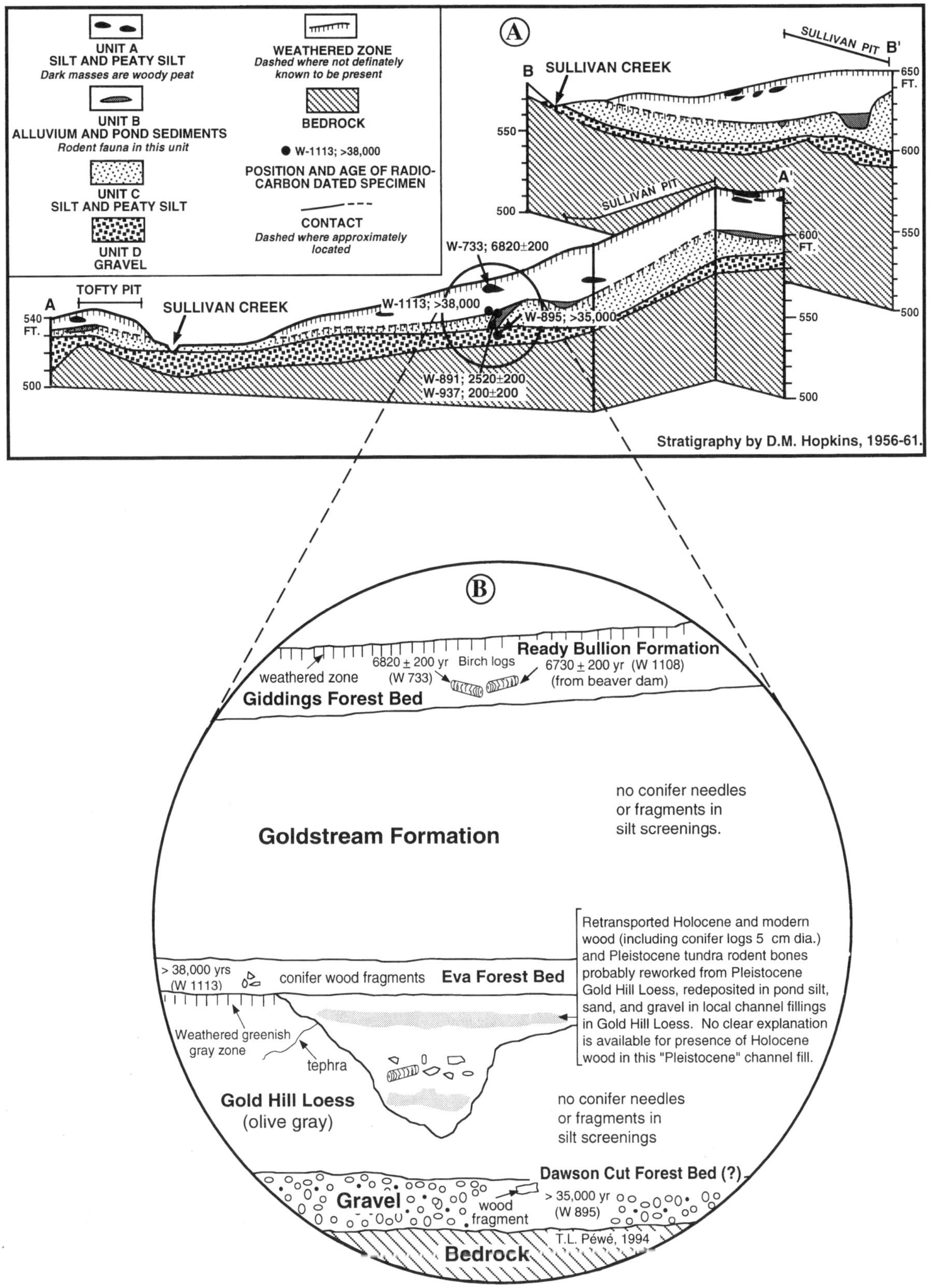
UNIT A
SILT AND PEATY SILT
Dark masses are woody peat
UNIT B
ALLUVIUM AND POND SEDIMENTS
Rodent fauna in this unit
UNIT C
SILT AND PEATY SILT
UNIT D
GRAVEL
WEATHERED ZONE
Dashed where not definately known to be present
BEDROCK
W-1113; >38,000
POSITION AND AGE OF RADIO-CARBON DATED SPECIMEN
CONTACT
Dashed where approximately located
A
B
SULLIVAN CREEK
SULLIVAN PIT
B'
A'
TOFTY PIT
W-733; 6820±200
W-1113; >38,000
W-895; >35,000
W-891; 2520±200
W-937; 200±200
Stratigraphy by D.M. Hopkins, 1956-61.
B
Ready Bullion Formation
weathered zone
6820 ± 200 yr (W 733)
Birch logs
6730 ± 200 yr (W 1108) (from beaver dam)
Giddings Forest Bed
Goldstream Formation
no conifer needles or fragments in silt screenings.
Retransported Holocene and modern wood (including conifer logs 5 cm dia.) and Pleistocene tundra rodent bones probably reworked from Pleistocene Gold Hill Loess, redeposited in pond silt, sand, and gravel in local channel fillings in Gold Hill Loess. No clear explanation is available for presence of Holocene wood in this "Pleistocene" channel fill.
> 38,000 yrs (W 1113)
conifer wood fragments
Eva Forest Bed
Weathered greenish gray zone
tephra
Gold Hill Loess (olive gray)
no conifer needles or fragments in silt screenings
Dawson Cut Forest Bed (?)
Gravel
wood fragment
> 35,000 yr (W 895)
T.L. Péwé, 1994
Bedrock

Figure 21. The most prominant stratigraphic feature of the frozen late Cenozoic deposits exposed in mining cuts of the Yukon-Tanana Upland is the remarkably well-developed unconformable contact between the dark retransported loess of the Goldstream Formation (Wisconsin) and the underlying massive light colored (green) Gold Hill Loess (pre-Sangamon). The Interglaciation Eva Forest Bed (Sangamon) occurs locally at the unconformity from one end of the .5-km-long exposure to the other. At the extreme right, the frozen silt cliff has thawed and the stratigraphy is obscured. Panoramic view of the east wall of Eva Bench Mining Cut 100 m west of Eva Creek (Fig. 3), 14 km west of Fairbanks, Alaska. Wood and pipes on gravel floor of cut and on cut surface of Goldstream Formation at left are from placer gold mining operations (photographs 482, 483, and 484 by T. L. Péwé, September 24, 1949).

streams such as the Tanana and Yukon River (Fig. 1) cut down permitting tributary streams to cut into creek-valley fills of frozen loess. In addition, thawing of permafrost and melting of ground ice provided extra water to slice into the easily erodible Gold Hill Loess. Considerable loess was removed on hill slopes and much, if not most, of the Gold Hill Loess in creek-valley bottoms was removed (Figs. 4, 11; also see Péwé, 1965a, figs. 1-8, 1-10). On lower slopes gullying of loess permitted giant sections of the sediment to undergo block slumping. Tilting of tephra layers outline the slump blocks (Figs. 8, 11a, 12, 13, 22; also see Péwé, 1952a, fig. 42, 1975b, fig. 7, p. 14).

Perhaps most of the permafrost thawed, as indicated by the absence of ice wedges (a few ice wedge casts exist), pingos, and frozen mammal carcasses in the now refrozen Gold Hill Loess and older formations. These features are, however, present in the overlying deposits of Wisconsin age. As the permafrost thawed, the iron coating on minerals in the upper part of the frozen fresh Gold Hill Loess was reduced to a green color, in contrast to the black, buff, or tan color of the unreduced overlying sediments. The thickness of the frozen loess with green color is approximately 10 to 20 m (frontispiece). Refreezing in Wisconsin time has preserved the green color which is rapidly lost when the loess is exposed and thawed today.

Examined stratigraphic sections indicate that the forest bed formed after much of the thermokarst thawing, erosion, and slumping activity had ceased. Almost all sections show the Eva Forest Bed overlying an angular unconformity between the underlying Gold Hill Loess and the forest bed (Figs. 8, 9, 11, 12, 13, 22). However, many of the sections illustrate no forest bed at all at the unconformity (Fig. 20; see Péwé, 1965a, 1975b, fig. 7; Berger et al., 1996; S. J. Preece et. al., 1997, unpub. data), especially on upper slopes and tops of low hills, such as Gold Hill and College Hill (University of Alaska). The forest bed could have been locally destroyed as some erosion continued. Microfaulting probably occurred throughout the interglaciation, such as shown on upper Eva Creek (Fig. 14). Also, because of continued loess removal as it fell on some sharp loess ridges on upper slopes and near steep bluffs, Gold Hill Loess may not have been buried and crops out at the surface today (Fig. 4; Péwé, 1992; T. L. Péwé, 1997, unpub. data), such as at the Halfway House exposure, Birch Hill, and others, and younger formations are not present.

Remnants of the Eva Forest Bed are rare under the Goldstream Formation where it overlies Fox Gravel in creek-valley bottoms (Figs. 4, 11; see Péwé, 1965a, figs. 1-8, 1-10). As small streams flowing on top of the gravel moved back and forth across the narrow valley floors, constantly reworking the top of the gravel, they probably removed accumulating deposits. One exception is a buried spruce stump "bed" (Fig. 23) on top of Fox Gravel underlying the Goldstream Formation on Amy Creek, a small stream near Livengood (Fig. 3, Table 1).

Relation of tephra layers in the upper Gold Hill Loess to Eva Forest Bed. Differential preservation of many distinct tephra layers in the upper Gold Hill Loess provides strong evidence for differential erosion of this loess before the development of the Eva Forest Bed. Here we outline the occurrence of these tephra beds and the history of their identification, while further comments on this stratigraphic matter are provided below in our discussion of geochronology.

Almost every stratigraphic section in the Yukon-Tanana Upland exposing upper Gold Hill Loess exhibits one or more tephra layers 0.5 to 15 cm thick (Figs. 4, 8, 9, 11, 12, 13, 14). Since the late 1940s, samples have been collected by Péwé from Gold Hill Loess tephra layers for use in correlating between stratigraphic exposures (T. L. Péwé, field notes). Two or three different tephra beds were recognized in the beginning of this work, but by the 1970s tens of different tephra were identified and correlated from these samples by work of J. A. Westgate and his associates. With development of the isothermal-plateau fission-track method (Westgate, 1988, 1989; Wagner and Van den haute, 1992), these fine-grained tephra beds could be dated accurately and precisely. In addition, they are readily distinguished from one another by their stratigraphic context and physicochemical attributes. Thus, these tephra beds could be used as key marker horizons, which

Figure 12. Stratigraphic sketch of unconsolidated Quaternary sediments in mining exposures at the Tofty Placer District near Manley Hot Springs in western Yukon-Tanana Upland, east-central Alaska, 120 km west of Fairbanks: (A) Pioneer stratigraphic, paleontologic, and sedimentologic interpretations, with radiocarbon dates of exposures by D. M. Hopkins, 1956–61 (see Repenning et al., 1964). (B) Stratigraphic interpretation with terminology by Péwé based on clear, detailed observations by Hopkins and on reconnaissance field observations by Péwé, 1949 and 1959, in the Tofty area, and extrapolation from similar mining exposures near Fairbanks by Péwé, 1946 to 1993. Tephra location from Péwé field notes, July 3, 1959. Sketch (A) modified from Repenning et al. (1964) and published with permission of *Arctic*.

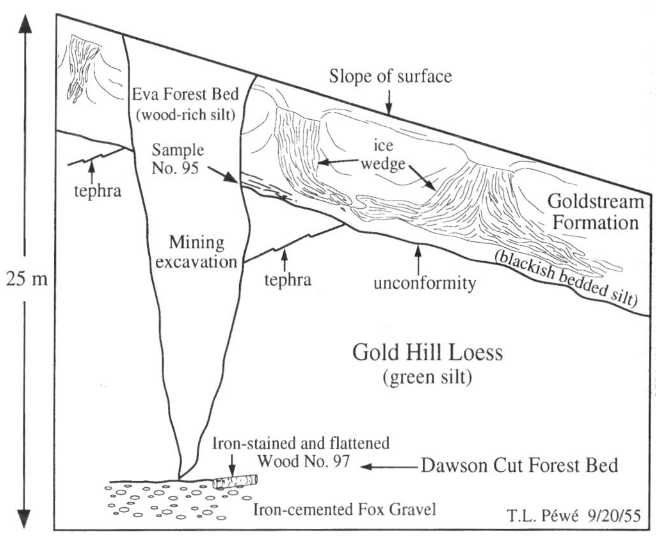

Figure 13. Sketch of perennially frozen loess and gravel exposed at east wall of Sheep Creek Cut 10 km northwest of Fairbanks, Alaska. Wood-rich silt (sample no. 95) of Eva Forest Bed crops out at base of Goldstream Formation unconformably overlying Gold Hill loess and tephra layers. Down-bowed silt layers over flat-topped Wisconsin ice wedges indicate early Holocene warm interval (Péwé, 1965a).

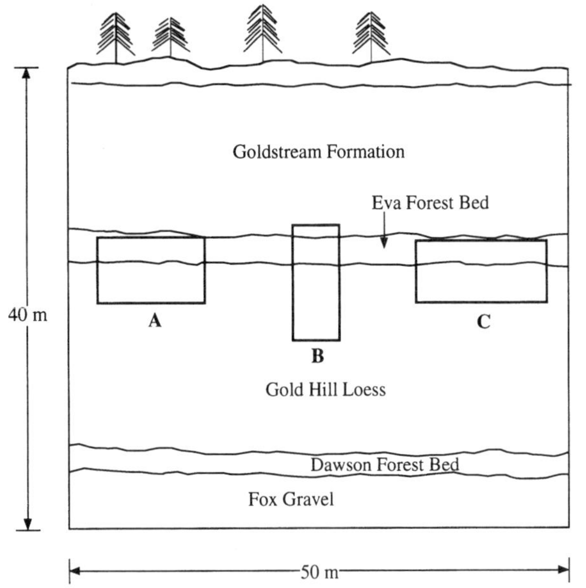

Index map of sections A, B, and C on upper Eva Creek mining cut.

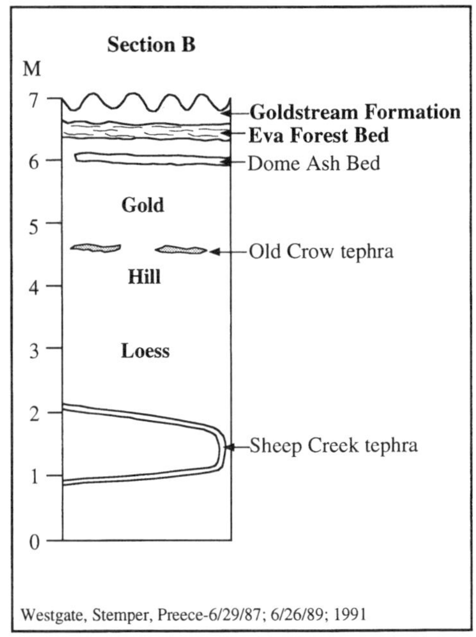

Section C

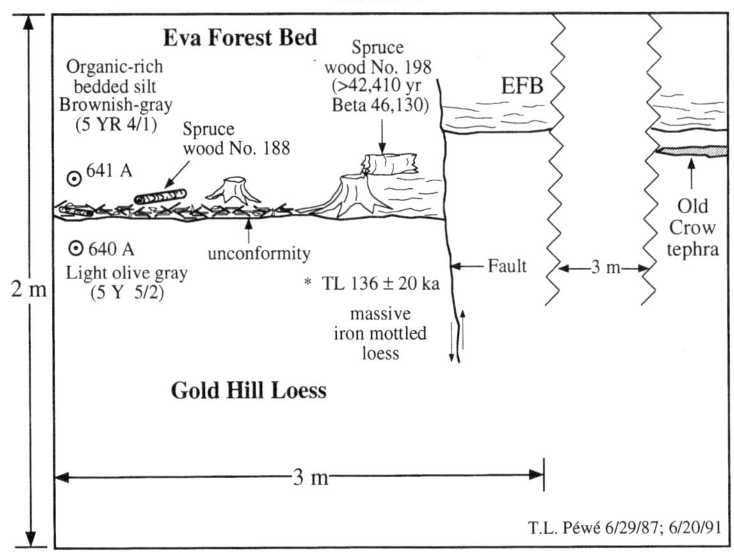

Figure 14 (on this and facing page). Three stratigraphic sections (A, B, and C) on west wall of upper Eva Creek mining cut, 14 km west of Fairbanks, Alaska, showing lower part of the Eva Forest Bed unconformably overlying upper Gold Hill Loess and enclosed tephra.

Section A. Large, 32 cm diameter spruce log of Eva Forest Bed protruding 1.5 m from silt cliff (sawed off by Péwé in 1987; see Fig. 16). Log overlies burnt fragments of birch and spruce (Table 1). Radiocarbon dating of log (wood no. 187) by liquid scintillation counting gives age greater than 70,000 yrs (A. Long, University of Arizona, January 4, 1993, oral commun.). Old crow tephra identified by S. J. Preece, 1991.

Section B. Seven-meter deep vertical trench dug by J. A. Westgate and associates exposing bedded silt of Goldstream Formation and Eva Forest Bed overlying upper Gold Hill Loess and interbedded tephra. Tephra identified by J. A. Westgate, B. Stemper, and S. J. Preece. Sheep Creek tephra overturned by solifluction. Age of tephra: Old Crow tephra, 140 ± 10 ka (fission track) (Westgate et al., 1990); Sheep Creek tephra, 190 ± 20 ka (TL) (Berger et al., 1996). These dates were not determined from tephra or loess of this locality. (Section from Preece, 1991.)

Section C. Spruce stumps and logs at base of Eva Forest Bed on unconformity between bedded organic-rich silt and underlying massive Gold Hill Loess (Table 1). Circle symbol is location of sediment sample for size-grade analyses (Fig. 15). Numbers by color description are Munsell color code. Old Crow tephra collected by T. L. Péwé and R. Reger in 1981 and identified by J. A. Westgate. Thermoluminescence date of 136 ± 20 ka on Gold Hill Loess by G. W. Berger (July 2, 1995, written commun.).

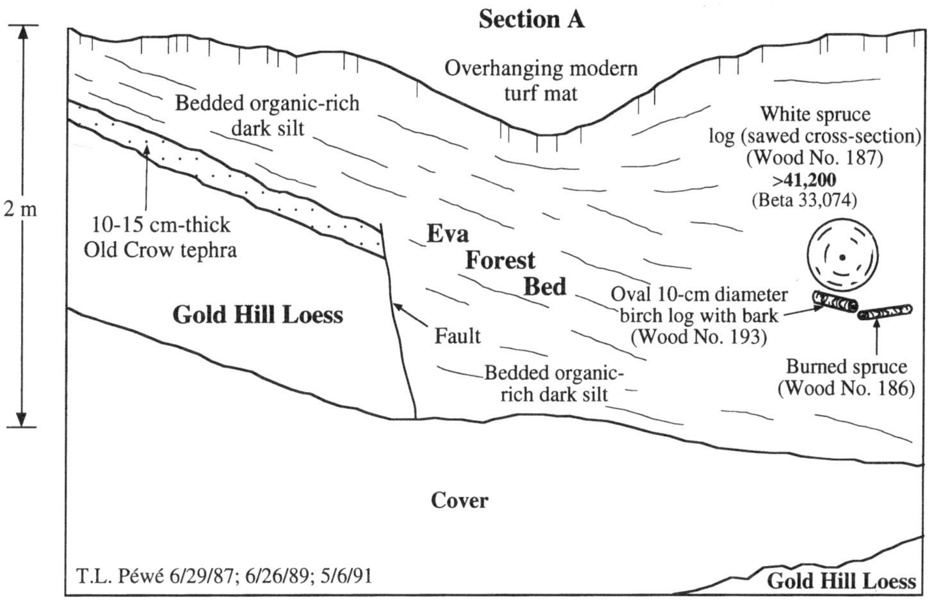

greatly facilitated understanding of the loess stratigraphy. There are at least 34 different identified tephra beds now known in the Gold Hill Loess (Péwé, in press). Preece and Westgate have identified and correlated 20 of the 34 different tephra beds by major and trace element geochemistry of glass shards, petrography, and stratigraphic sequence (S. J. Preece et al., 1997, unpub. data).

Six different tephra beds from the upper part of the Gold Hill Loess have been identified and characterized by the work of Westgate and associates since 1971. From the oldest to youngest the tephra are Sheep Creek, Old Crow, Halfway House, SD, and Variegated tephra. Dome Ash Bed (Péwé, 1975b) is younger than Old Crow tephra, but its age in relation to the younger three is unknown (Table 4). All but the SD tephra occur in multiple localities. A TL date of >207 ± 40 ka on College Hill tephra (Table 4; G. W. Berger, 1992–1995, unpub. data) indicates that it is older than the previously mentioned six tephra beds. Preliminary examination by Westgate indicates that on the basis of shard morphology the College Hill tephra does not correlate to the Sheep Creek tephra. Any of these seven tephra may lie stratigraphically at 1 to 5 m below the unconformity (Table 4; Fig. 4) between the Gold Hill Loess and the Goldstream Formation or the Eva Forest Bed (Figs. 4, 8, 9, 10, 14, 22). Such unconformity-proximal preservation likely arose because of block slumping and/or differing amounts of erosion of the Gold Hill Loess having tephra beds prior to the deposition of the Eva Forest Bed and the Goldstream Formation.

Eleven locations in the Yukon-Tanana Upland (Figs. 2, 3) where one or more of these seven mentioned tephra are at or just under the unconformity (Péwé, field notes) are indicated in Table 5. There would be 15 such occurrences if we count multiple sections in the same mining cut (upper Eva Creek, 3; Dawson Cut, 2; Gold Hill, 2). In many other exposures a tephra bed lies just below the unconformity, but the collected tephra sample has not been identified (see, e.g., Dome Creek, Fig. 9, Ester Island, Fig. 22, and Tofty, Fig. 12). Differential erosion of Gold Hill Loess with enclosed organic, slightly clay-rich loess (paleosols?) could also account for the different position of the Old Crow tephra in regard to these "soil" zones.

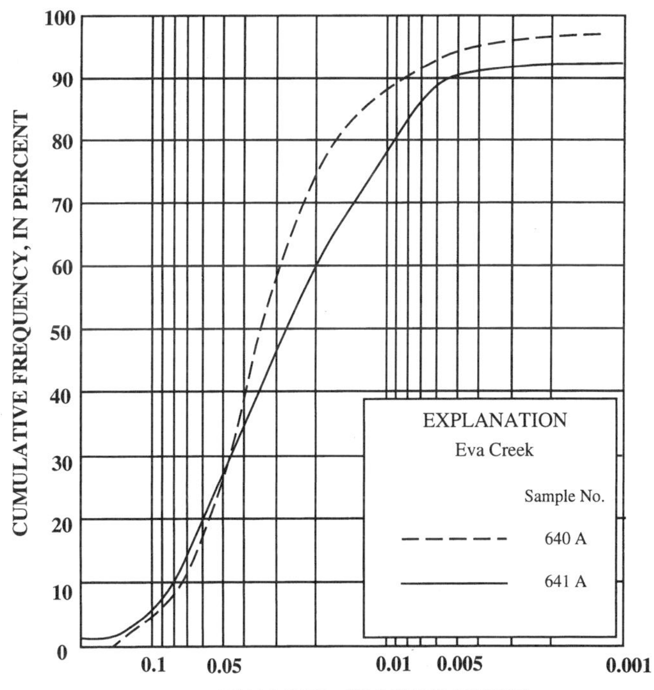

Figure 15. Cumulative-frequency grain-size curves of Gold Hill Loess (640 A) and retransported eolian silt from the Eva Forest Bed (641 A). Eva Creek mining cut 16 km west of Fairbanks, Alaska (see Figs. 3 and 14C). Samples collected by T. L. Péwé and analyzed at Department of Geology, University of Toronto, Canada.

Figure 16. Sawing off 1 m-long section of 32 cm diameter white spruce log (sample no. 187) protruding from perennially frozen 125,000 yr old Eva Forest Bed on west wall of upper Eva Creek mining exposure 16 km west of Fairbanks, Alaska. Tree has 144 rings up to 1 mm-wide (See Figs. 3, 14A, and cover photograph; also Table 1). (T. L. Péwé photograph PK 28,461, June 29, 1987.)

Some workers have attempted to use the variable thicknesses of loess preserved between the Old Crow tephra and the Eva Forest Bed, or the associated unconformity, as an argument for more than one "Old Crow tephra," or as an argument for different starting ages of the past interglaciation, or different rates of wind intensities and different rates of loess deposition. (Begét and Hawkins, 1989; Begét, 1990; Begét et al., 1991; Hamilton, 1993; Hamilton and Bringham-Grette, 1991). This interpretation of the loess stratigraphy appears to be overly simplistic, for it ignores the probable

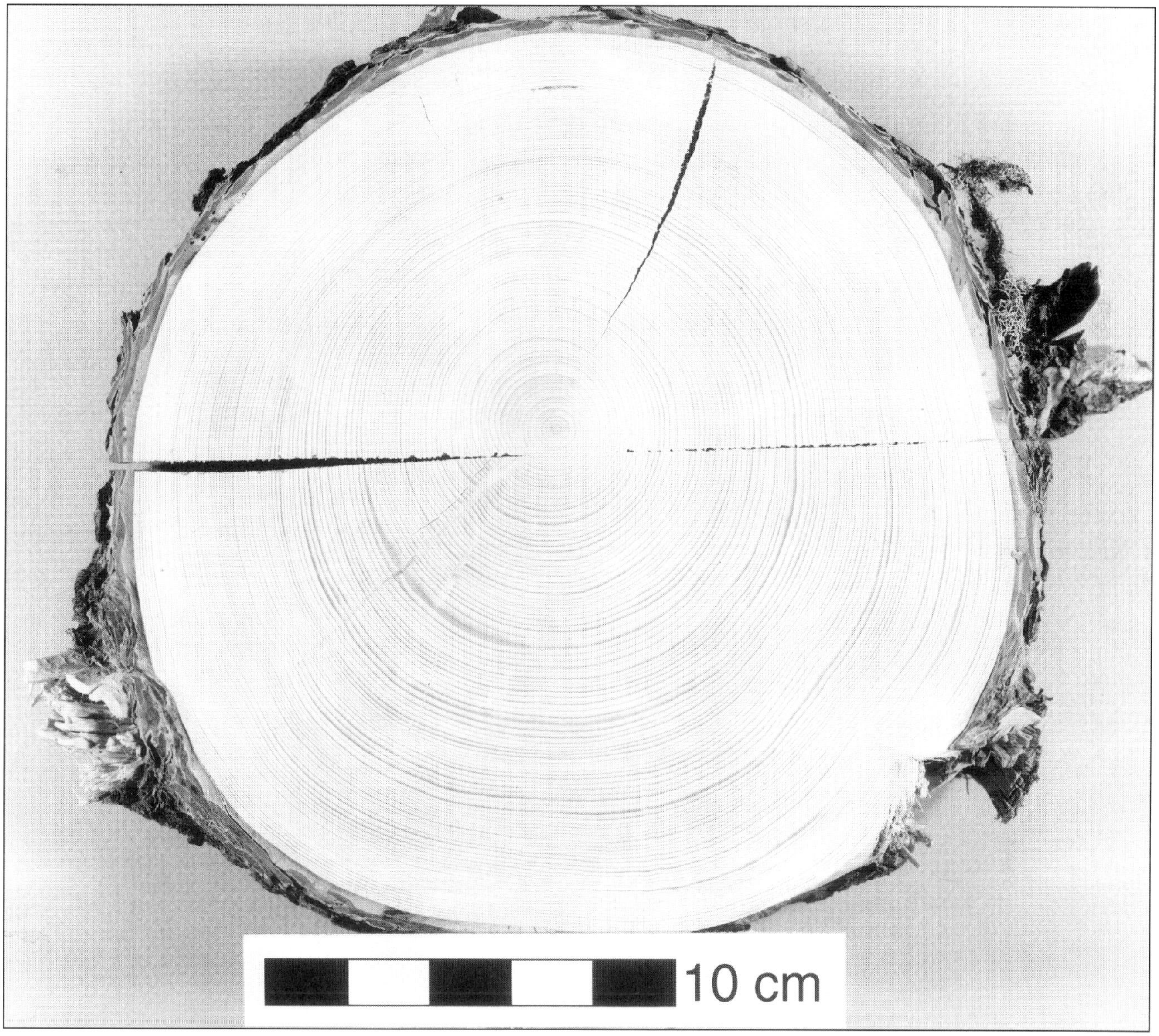

Figure 17. Stem disk of modern, mature white spruce (*Picea glauca*) (center date, 1877) which grew at elevation of 260 m on south-facing slope, Bonanza Creek Experimental Forest 30 km west of Fairbanks, Alaska. Tree blown down in 1990. Section collected and cut by Les Viereck, U.S. Forest Service, 1990. Tree rings dated by Peter Brown. Horizontal cut across stem disk made to obtain sample for $^{13}C/^{12}C$ study. Péwé sample no. M2 (Table 2). (Photograph no. 3894-H by D. H. Ball, Arizona State University, Tempe, May 4, 1993.)

occurrence of differential erosion of the upper Gold Hill Loess at different sites.

There are three localities where the Old Crow tephra occurs in well-exposed Gold Hill Loess, but there is no direct relationship to the Gold Hill Loess–Goldstream Formation unconformity or the Eva Forest Bed at these localities because the younger formations have been removed. Two of these localities have been studied in some detail. Based on our findings in this report we differ from some previously published interpretations of one locality: Halfway House.

Halfway House. Along the Parks Highway, 35 km west of Fairbanks, is an exposure that we believe to be composed entirely of Gold Hill Loess 12 m thick with an imprint of a Holocene soil on the top (T. L. Péwé, field notes of 1983, 1987, 1991, 1993). Four tephra beds occur stratigraphically from bottom to top: Old Crow, Halfway House, SD, and the Variegated (Preece, 1991). Under and just above the Old Crow tephra, and elsewhere in the section, are darker layers of loess with a higher organic (?) content. These have been interpreted as significant interglacial or interstadial paleosols. The stratigraphically lowest "paleosol" is

TABLE 2. BASIC DATA FOR MODERN WOOD SAMPLES FROM NEAR FAIRBANKS, ALASKA

Sample Number	Location	Date	Specimen	Description	Tree Rings	Identified	Discussion*	Comments
M1	Tanana River Floodplain 30 km west of Fairbanks	1987	White spruce	Six 5 cm thick sections		R. Reger, Péwé, UAZTRL		Island in Tanana River. Collected by A. Youngblood. Donated by R. Reger. Elevation 120 m
M2	Bonanza Creek Experimental Forest; 30 km west of Fairbanks	1990	White spruce *(Picea glauca)*	10 cm thick section; 40 cm diameter	113	L. Viereck, U.S. Forest Service	Section 3 m above base; tree blew down December 1990; mature tree, 30 m height	Collected by L. Viereck. South-facing slope. Elevation 290 m; Figure 17, this report
M3	Bonanza Creek Experimental Forest; 30 km west of Fairbanks	1990	White spruce *(Picea glauca)*	10 cm thick section; 40 cm diameter		L. Viereck, U.S. Forest Service	Section 10 m above base; tree blew down December 1990; mature tree, 30 m height	Collected by L. Viereck. South-facing slope. Elevation 290 m
M4	Bonanza Creek Experimental Forest; 30 km west of Fairbanks	1983	White spruce *(Picea glauca)*	10 cm thick section; 40 cm diameter		L. Viereck, U.S. Forest Service	Section 1 m above base; killed in 1983 Rosie Creek fire	Collected by L. Viereck. South-facing slope. Elevation 290 m
M5	Bonanza Creek Experimental Forest; 30 km west of Fairbanks	1983	White spruce *(Picea glauca)*	10 cm thick section; 40 cm diameter		L. Viereck, U.S. Forest Service	Different tree from no. M4; Section 1 m above base. Killed in 1983 Rosie Creek fire	Collected by L. Viereck. South-facing slope. Elevation 290 m
M6	Fairbanks	6/22/91	Cottonwood *(Populus balsamifera)*	Tree section 60 cm above base; 24 cm diameter	22	Péwé, UAZTRL	Tree cut down 6/21/91 on Chena River floodplain 250 m south of river	Depth to permafrost >2 m. Collected by Péwé. Elevation 140 m
M7	Goldstream Creek Valley bottom; 7 km east of Fox on gravel tailing piles	6/22/91	Cottonwood *(Populus balsamifera)*	Log section, 40 cm long, 14 cm diameter; beaver chewed; 40 cm diameter at base	30?	Péwé, UAZTRL	Tree felled by beaver about 10 years earlier	Depth to permafrost 1 m; Collected by Péwé. Elevation 250 m
M8	Chatanika River floodplain; 20 km north of Fairbanks	6/23/91	Diamond willow *(Salix)*	3 cm diameter; 4 cm long stick		Péwé, UAZTRL	Wood cut 8 years earlier	Donated by J. Murphy. Depth to permafrost about .5 m
M9	Goldstream Creek Valley bottom; 7 km NE of Fairbanks Sec. 15 T2N R1E	6/28/91	Cottonwood *(Populus balsamifera)*	23 cm diameter log; beaver chewed on both ends		Péwé, UAZTRL	Modern beaver dam	Depth to permafrost 1 m; collected by Péwé. Elevation 300 m
M10	Goldstream Creek Valley bottom; 7 km NE of Fairbanks Sec 15 T2N R1E	6/28/91	Birch or aspen	2 cm diameter stick		Péwé, UAZTRL	Modern beaver dam	Depth to permafrost 1 m; collected by Péwé. Elevation 300 m

*Microfilm of notebooks on file, U.S. Geological Survey, Denver, CO 80225.
UAZTRL = University of Arizona Laboratory for Tree Ring Research, Tucson, AZ 85721. Peter Brown identified specimens and determined number of tree rings.

Figure 18. Stem disk of spruce (*Picea* sp.) cut from log preserved in permafrost in Eva Forest Bed. Extremely narrow rings characteristic of black spruce or white spruce growing under difficult environmental conditions. Tree was 179 years old and grew at elevation of 250 m on south-facing slope. Collected by Péwé in 1991 from mining exposure at Eva Creek, 14 m west of Fairbanks, Alaska. Stem disk prepared and tree rings counted by Peter Brown. Péwé sample no. 198 (Table 1). (Photograph no. 4013-H by D. H. Ball, Arizona State University, Tempe, June 2, 1994.)

thought to represent the last interglaciation (Sangamon) (Westgate et al., 1983; Begét and Hawkins, 1989; Begét, 1990; Begét et al., 1991; Hamilton and Brigham-Grette, 1991; and Hamilton, 1993). We believe they are local organic, slightly clay-rich soil layers that may or may not indicate minor weathering intervals.

We believe the stratigraphically lowest dark loess layer (paleosol) does not represent the last interglaciation (125,000 years old) for the following reasons: (1) The Old Crow tephra occurs within the paleosol and the tephra is 140 ± 10 ka. (2) Stratigraphically above the Old Crow tephra, and extending to within 5 m of the top of the 12 m thick loess exposure, occur three tephra in the following order, bottom to top: Halfway House, SD, and Variegated. The Old Crow, Halfway House, and Variegated tephra occur in other localities in the Fairbanks area (Table 5) near the top of the Gold Hill Loess just under the unconformity between the Gold Hill Loess and the Goldstream Formation. The unconformity has been demonstrated to be of Eva Interglaciation time (Sangamon). (3) The tephra layers trend laterally to the edge of the hill showing that the hillock has been carved entirely of Gold Hill Loess. (4) Péwé has studied details of the loess here since 1983 (T. L. Péwé, 1997, unpub. data) and 15 mechanical analyses and field observations at the site support the concept that the outcrop exposure is Gold Hill Loess. The tan to light olive gray, massive loess throughout the section has essentially the same color, texture, and absence of bedding as does Gold Hill Loess in the Fairbanks area. The silt here and at Birch Hill, noted below, is well sorted with 75 to 95% of the particles falling between 0.063 mm and 0.002 mm in diameter. Many young anomalous radiocarbon dates are available from the loess in the surface zone influenced by Holocene weathering and are part of a study underway (T. L. Péwé, unpub. data). (5) In addition to the beliefs of the authors, R. D. Reger, formerly of the Alaska Division of Geology and Geophysics, who has examined loess of the Fairbanks area and elsewhere in Alaska from the early 1960s to 1994, also believes that the Halfway House exposure is entirely of Gold Hill Loess.

TABLE 3. BASIC DATA FOR WOOD SAMPLES FROM GIDDINGS FOREST BED (HOLOCENE) NEAR FAIRBANKS, ALASKA

Sample Number	Location	Date	Specimen	Description	Tree Rings	Age ^{14}C	Identified	Discussion and Stratigraphy*	Comments
2	Wilbur Creek	6/2/51	Root	10 cm diameter root	70		Spruce (UAZTRL)†	Base of Ready Bullion Formation(?)	On fallen block
4	Lower Eva Creek	6/5/51	Stumps and logs	Stumps and logs 10 cm diameter		3,750 (L117H)	Spruce and birch (Péwé)	Base of Ready Bullion Formation; 2.8 m below surface	See Figure 11C, this report, specimen not available
7a b c	Fairbanks Creek	6/6/51	Three 2 to 5 cm diameter sticks	Beaver-chewed wood sticks (3)	13+1(P)† a 37+(W)† b 23 (W) c	9,350 + 80 (Beta 58,408) Sample 7b (13,600 ± 600, 1952 solid carbon date L117I)	*Populus* (1) Diamond willow (2) (Péwé, UAZTRL)	From beaver dam at base, silt fill of beaver lake	Center of valley; transverse to valley. Buried by later organic silt with fresh water snails; see Figure 48, Péwé, 1952a
6	Gold Hill	6/5/51	Stump	15 cm diameter upright stump	83	4,020 ± 200 (W183)	Birch (Péwé, UAZTRL)	Base of Engineer Loess 2 m below surface	Péwé, 1965b, p. 20; Péwé and Reger, 1983
8	Fairbanks Creek	6/6/51	Log	15 cm diameter birch log	79		Birch (Péwé)	Unconformably overlying pingo ice, base of Ready Bullion Formation 2 m below surface	
10	Dawson Cut	6/9/51	Log	Well-preserved 8 cm log of beaver dam	45	7,280 ± 80 (Beta 50,683)	Spruce (UAZTRL)	Above top of Goldstream Formation (8 m below surface and 30 m above gravel) in beaver dam. Dam is in lower part of Ready Bullion Formation	Both birch and spruce present in dam Figure 10, this report
17	Gold Hill	6/1/52	Root	20 cm long, 4 cm diameter root	33		Spruce (Péwé, UAZTRL)	At base of Engineer Loess 1.2 m below surface	Upright stump and logs in forest bed
....	Dome Creek	6/8/52	Log	15 cm diameter beaver-chewed log	56	8,800 ± 70 (Beta 50,685)	*Populus* (Péwé, UAZTRL)	Beaver dam in Ready Bullion Formation	Collected by Otto W. Geist. Narrow rings
30	Fairbanks Creek	6/10/52	Fragment of stump 18 cm long	Spruce; stump upright, approximate diameter 20 cm	50	6,040 ± 240 (W434)	Spruce (UAZTRL)	Stump at base of Ready Bullion Formation, 1.8 m below surface	
49	Dome Creek	8/16/54	Sticks	1 to 2 cm diameter sticks of *Populus* and diamond willow		9,650 ± 140 (Beta 69.379)	(Péwé, UAZTRL)	In beaver dam but not chewed	

TABLE 3. BASIC DATA FOR WOOD SAMPLES FROM GIDDINGS FOREST BED (HOLOCENE) NEAR FAIRBANKS, ALASKA (continued, p. 2)

Sample Number	Location	Date	Specimen	Description	Tree Rings	Age ^{14}C	Identified	Discussion and Stratigraphy*	Comments
106	Sheep Creek	8/14/56	Root	Fresh spruce root from overturned stump	62	5,940 ± 250 (W859)	Spruce (Péwé, UAZTRL)	Wood silt buried in Holocene gravel fan overlying Goldstream Formation	Right limit of cut. Birch with bark also present. A forest buried by torrential gravel deposit.
111	Sheep Creek	7/2/57	Stick	Beaver-chewed wood; 3 cm diameter stick	15	6,100 ± 80 (Beta 52,700)	*Populus* (Péwé, UAZTRL)	3 m below surface in alluvial silt of Goldstream Creek	In Goldstream Creek bank exposure at mouth of Sheep Creek Cut
185	Upper Eva Creek	6/27/87	Root(?) fragments	4 cm diameter fragments	Broken	1,430 ± 60 (Beta 22,110)	Spruce (Péwé, UAZTRL)	In upper part of Ready Bullion Formation 1.9 m below surface	
199	Upper Eva Creek	6/26/91	Beaver-chewed sticks	Sticks without bark .5 to 5 cm in diameter	77	8,940 ± 80 (Beta 46,215)	Spruce (Péwé, UAZTRL)	In beaver dam; near base of Ready Bullion Formation	On fallen block of organic, bedded retransported loess. Associated with *populus* specimens. Spruce rare.
200	Upper Eva Creek	6/26/91	Beaver-chewed sticks	5 cm diameter sticks	33+	8,780 ± 100 B.P. (Beta 48,788)	Diamond willow (Péwé)	In beaver dam; near base of Ready Bullion Formation	On fallen block of organic, bedded retransported loess

*All localities have stratigraphic diagrams in notebooks.

†UAZTRL = University of Arizona Laboratory for Tree Ring Research, Tucson, AZ 85721. Peter Brown identified specimens and determined number of tree rings. Samples collected by T. L. Péwé. P = poplar; W = willow

TABLE 4. STRATIGRAPHIC POSITION OF TEPHRA BEDS IN THE UPPER PART OF GOLD HILL LOESS NEAR FAIRBANKS, ALASKA*

Variegated tephra	
SD tephra	Dome Ash Bed
Halfway House tephra	
Old Crow tephra	
Sheep Creek tephra	
College Hill tephra	

*Oldest at bottom; Dome Ash Bed younger than Old Crow tephra but relation to the youngest three tephra is unknown. See Figure 4.

Magnetic susceptibility measurements of loess samples from this section by Begét and coworkers suggested that the deposition of the loess occurred from about 250 ka or 150 ka to the present (Begét and Hawkins, 1989; Begét, 1990; Begét et al., 1990). From sediment and tephra examinations mentioned above, we believe that the loess of the whole section is older than 125 ka, and the magnetic susceptibility curve of Begét represents only the upper part of the Gold Hill Loess.

Birch Hill, Fairbanks. A road cut on a south-facing, steep bluff (Fig. 3) traverses a loess ridge composed entirely of Gold Hill Loess and exposes a horizontal 12.5 cm thick layer of Old Crow tephra that crops out intermittently for 60 m. The tephra was identified by Westgate (written communication, 1985). There are no "paleosols" just above or below the tephra. The exposure with tephra was noted and photographed by Péwé (1952a, figs. 21, 23; and 1955, pl. 2, fig. 2; pl. 3, fig. 2) and studied in detail by Péwé (1983–1995, field notes). The TL ages on loess of 128,000 ± 22,000 years just above the Old Crow tephra and 144,000 ± 22,000 years just below were obtained by Berger et al. (1996).

Mile 293, Richardson Highway. Eighty km SW of Fairbanks, a layer of Old Crow tephra 2 to 20 cm thick in Gold Hill Loess crops out in a road cut of the Richardson Highway (Péwé, 1989; Péwé and Reger, 1989, p. 20). Tephra was identified by Westgate. No paleosol was noted just above or under the tephra layer.

Figure 19. Part of 16-cm-diameter horizontal white spruce log (sample no. 198; Table 1) from Eva Forest Bed, Eva Creek, 14 km west of Fairbanks, Alaska. The debarked surface of the log is scored by well-developed, multiple galleries created by ancient spruce beetles (Scolytidae). The original egg galleries are constructed parallel to the grain of the wood. The galleries at right angles to the longitudinal galleries are created by the newly hatched larvae. (Photograph PK 29,737, by T. L. Péwé, June 23, 1991.)

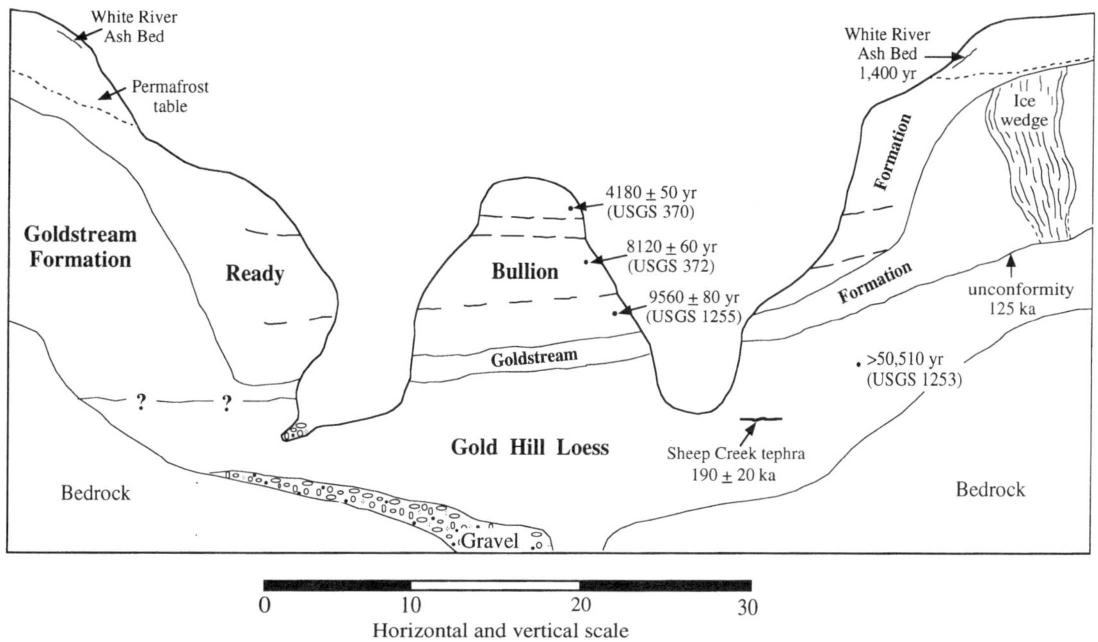

Figure 20. Quaternary stratigraphy of perennially frozen sediments at Lost Chicken Creek, Alaska, as interpreted by T. L. Péwé from Porter (1988, fig. 2) based on 1955 field examination by Péwé and extrapolation from similar deposits in the Fairbanks area, Alaska, by Péwé (1946–1993). Sheep Creek tephra identification by J. A. Westgate (Porter, 1988, p. 304). Age of Sheep Creek tephra by thermoluminescence dating of loess above and below the tephra in the Fairbanks area (Berger et al., 1996). Modified from Porter (1988) and published with permission of *Arctic*.

Age

The age of the interglacial Eva Forest Bed is one of the most critical aspects of its interpretation. Efforts to determine the age of the Eva Forest Bed have covered the past 50 years and span the time when Quaternary geological events were dated mainly by reference to ordinal sequence and long-established models of climatic change to the modern era of application of newly developed numeric dating methods.

Early stratigraphic interpretation. In the late 1940s and 1950s there were many active mining exposures of frozen sediments in the Yukon-Tanana Upland that changed daily in the summer and provided fresh, three-dimensional views of the Quaternary stratigraphy. These exposures created a type of stratigraphic fence diagram (see, e.g., Péwé, 1952a, figs. 33, 34, 35, 37), in contrast to a single thawed face of loess that must be trenched today to permit even a limited view.

These multiple exposures of loess layers of different colors and structures permitted the careful construction of major units (frontispiece, Fig. 21—see center foldout). A young, organic-rich loess deposit with ice wedges, pingos, and frozen animal carcasses unconformably overlay an older massive loess. The older loess, although frozen, contained no ice wedges, pingos, nor frozen carcasses. The Eva Forest Bed was exposed locally at the unconformity between these two loess formations (Fig. 4). The younger loess was interpreted to represent Wisconsin glacial time and the older loess, pre-Wisconsin (Illinoian or older) glacial time. The Eva Forest Bed was considered to be interglacial or Sangamon in age (Péwé, 1952a, 1952b, 1954, 1955, 1957, 1958, 1965c; Péwé and Hopkins, 1967; Matthews, 1968, 1970).

This interpretation of the loess stratigraphy and age, was supplemented by relating these deposits to the glacial sequence on the north side of the Alaska Range, only 80 km away at the Delta River (Fig. 2). A sequence of three major glacial advances was noted in 1949 (T. L. Péwé, field notes) and later related to the Wisconsin and pre-Wisconsin loess blankets in the Yukon-Tanana Upland, especially near Fairbanks (Péwé, 1952c, 1957, 1958, 1965a, 1965b; Péwé et al., 1953; Péwé and Holmes, 1964; Holmes and Péwé, 1965; Coulter et. al., 1965; Blackwell, 1965).

Standard, stratigraphic dating of the Eva Forest Bed in the 1950s and 1960s, using sequences of both periglacial and glacial deposits, indicated a Sangamon interglacial age (Oxygen Isotope Substage 5e). Since the early 1950s radiocarbon dating has been applied to selected appropriate materials. Within the last decade, modern geochronometric techniques (radiocarbon, fission-track, thermoluminescence, paleomagnetic) as well as geochemical fingerprinting of tephric glass and minerals have been applied to various deposits associated with the Eva Forest Bed.

Radiocarbon dating. Beginning in 1949, samples of wood and other organic matter were collected from the frozen loess deposits in central Alaska for radiocarbon dating (Péwé, 1975b, table 4; Figs. 10, 11c, 22). Holocene and late Wisconsin deposits are well dated (Table 3). Samples from the early Goldstream Formation and Eva Forest Bed consistently yielded ages greater than 30

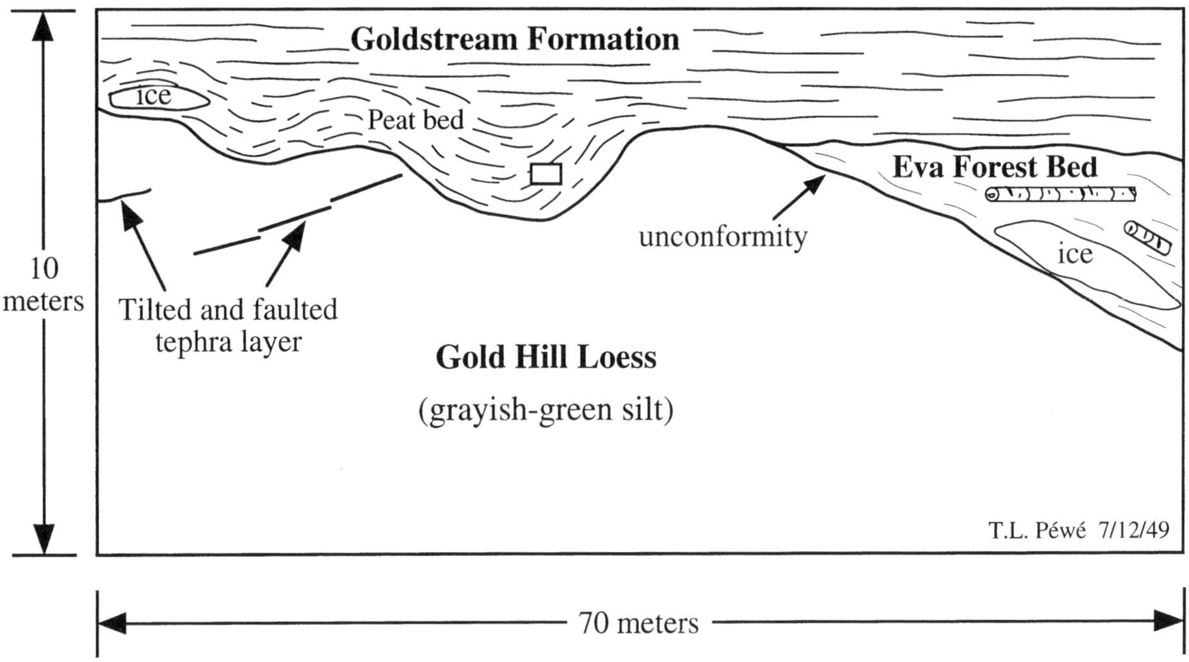

Figure 22. Diagrammatic sketch of the north-facing wall of Ester Island facing Ester Creek (Fig. 3) 14 km west of Fairbanks, Alaska. Goldstream Formation, with peat bed, and the Eva Forest Bed overlie grayish-green silt of Gold Hill Loess with a sharp but undulating contact. The stratigraphy is the same as the cut at lower Eva Creek, 500 m north (Fig. 11). Small rectangle indicates site of pollen sample from peat bed. Pollen analysis indicates treeless environment. Eva Forest Bed was probably more extensive, but removed by erosion prior to deposition of Goldstream Formation (from Péwé, 1952a, fig. 39).

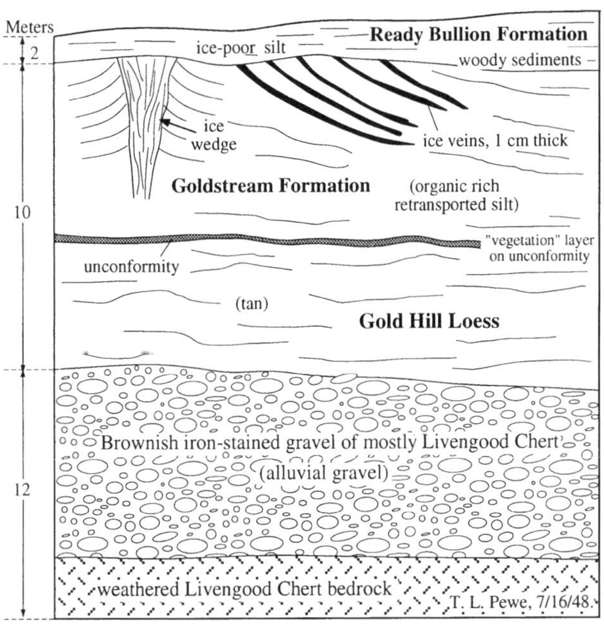

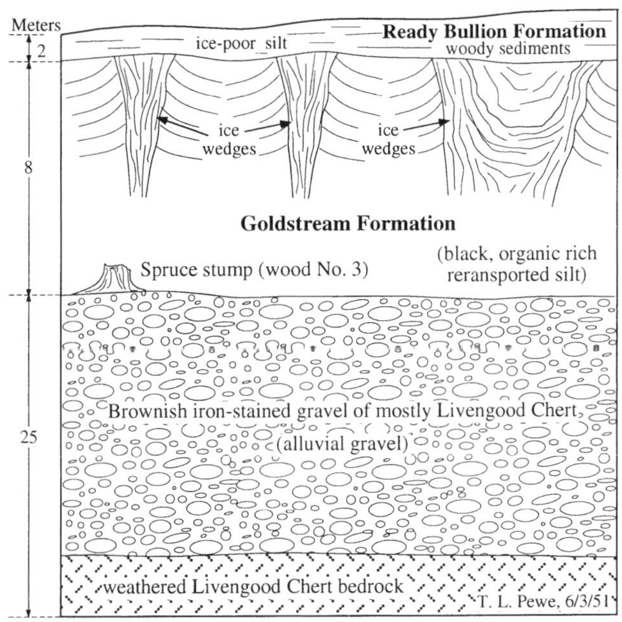

Figure 23. Diagrammatic sketches of perennially frozen late Cenozoic loess and gravel exposed over bedrock in 1948 and 1951 at Amy Creek, 100 km north of Fairbanks, Alaska. Eva Forest Bed illustrated as "vegetation" layer on unconformity at top of Gold Hill Loess in the 1948 location and as spruce stump "bed" in the 1951 location.

TABLE 5. LOCATIONS WHERE ONE OR MORE TEPHRA IN UPPER GOLD HILL LOESS IS AT OR JUST UNDER THE UNCONFORMITY BETWEEN THE GOLDSTREAM FORMATION OR EVA FOREST BED AND THE UNDERLYING GOLD HILL LOESS*

Locality	Tephra	Exposures at Locality	Figure	Tephra Under Eva Forest Bed
West Dawson Cut	Halfway House	1		
Dawson Cut	Old Crow Sheep Creek	2	10†	Old Crow
Gold Hill I	Variegated Halfway House	1	24†	
Gold Hill II	Old Crow	1	Péwé and Reger, 1989, Fig. 7; Péwé, in press, Fig. 4	Old Crow
Sheep Creek Cut	Old Crow Sheep Creek	2	24†	
Lower Eva Creek	Dome Sheep Creek	1	Péwé, 1975b, Fig. 7	Dome
Upper Eva Creek	Dome Old Crow Sheep Creek	3	14A, B, C†	Dome Old Crow Sheep Creek
College Hill	College Hill	1	24†	
Lost Chicken Creek	Sheep Creek	1	20†	
Birch Creek	Old Crow	1	Edwards and McDowell, 1991, table 1	Old Crow
Palisades	Old Crow	1	Begét, et al., 1991, fig. 3	Old Crow

*See Figures 2, 3, 4.
†Figures in this report.

or 40 ka (Table 1; Figs. 10, 11C, 12, 14). An age of older than 56.9 ka was obtained from a sample of the Eva Forest Bed from Eva Creek (Matthews, 1970).

After these age estimates from the 1950s and 1960s were obtained, no significant new radiocarbon ages of wood from the Eva Forest Bed were obtained for 25 years. However, in the early 1990s, a sample of a well-preserved spruce log (Table 1, sample no. 187; Fig. 14B) from the Eva Forest Bed exposed at Eva Creek was used as a "dead" radiocarbon sample (>100 ka) at the Environmental Radioisotope Center, Department of Geosciences, The University of Arizona, Tucson, Arizona. There, radiocarbon dating of wood in the 50 to 70 ka range without isotopic enrichment is being accomplished using liquid scintillation (LS) detectors. Preliminary work suggested that sample no. 187 was about 70 ka old (Long and Kalin, 1992a, 1992b; McCormac et al., 1993), but further study by Long, however, indicated that the specimen is older than 70 ka (A. Long, Jan. 4, 1993, written communication) This new radiocarbon age further supports the possibility of an age of the last interglaciation (125 ka) for the Eva Forest Bed.

Tephrochronology. The dating of tephra beds in the loess has long been held as a possible key to an understanding of the loess chronology in central Alaska and the age of the Eva Forest Bed. For example, many samples of loess and tephra were submitted to the U.S. Geological Survey for analyses in 1950. A sample of Ester Ash Bed (Péwé, 1955, 1975b) from the lower Gold Hill Loess (Fig. 4) was submitted to a commercial laboratory (Geochron Laboratory) in 1972, and a bulk potassium-argon date of 60 Ma was obtained. Another sample of Ester Ash Bed collected by Péwé was submitted to Westgate and the glass shards were dated as greater than 450 ka (Naeser et al., 1982) by the bulk-glass fission-track method. This tephra bed was later redated by the isothermal-plateau fission-track method as 810 ka (Westgate et al., 1990), a date supported by paleomagnetic studies (Westgate et al., 1990).

This historical review illustrates the progress that has been made in dating geologically young tephra near Fairbanks. This improvement in geochronometric techniques is paramount in dating tephra layers in the Gold Hill Loess that are stratigraphically close to the Eva Forest Bed (Figs. 4, 8, 10, 11, 12, 14, 20, 22). The Old Crow tephra, the thickest and most widespread of tephra beds near the Eva Forest Bed, was the first such "young" tephra in the region to be dated by fission-track methods. An attempt to date the glass shards in a sample from the Yukon Territory was unsatisfactory because no spontaneous tracks were observed due to the small glass surface area scanned—a result of the very fine grained character of this sample. Naesar et al. (1982) concluded that the Old Crow tephra was unlikely to be older than 120 ka, and therefore, some scientists believed that the overlying Eva Forest Bed to be early Wisconsin in age and the associated unusual climate warming occurred at that time (Carter and Ager, 1989; Schweger, 1989; Schweger and Matthews, 1985).

Continued refinement by Westgate (1987, 1988, 1989) of the fission-track method for dating geologically young tephra beds permitted a careful dating of Old Crow tephra from coarser-grained samples from Halfway House and Holitna, Alaska, much nearer the tephra's source in the eastern Aleutian arc (Westgate et al., 1985). The weighted mean (six analyses) age for the Old Crow tephra is 140 ± 10 ka (Westgate et al., 1990). This date strongly supports the original hypothesis of a last interglaciation age (125 ka) of the Eva Forest Bed. A closer constraint on the age of the last interglaciation in central Alaska awaits the fission-track dating of the Dome Ash Bed, one of the youngest tephra beds in the upper Gold Hill Loess (Table 4).

Thermoluminescence. Because thermoluminescence (TL) sediment dating is a technique that has been in a state of development for 16 years (since Wintle and Huntley, 1980), and because some earth scientists mistakenly continue to treat equally both early development and later TL dates, we present a historical summary of the application of TL dating to loess in Alaska and the central United States. This summary shows that as technique development has progressed, some TL ages for a given deposit have increased. A current discussion of why many of the early-development TL ages are erroneously young is given by Berger (1994).

In the 1980s, the TL method of numeric dating of feldspars in loess and glass from volcanic ash was widely applied in attempts to establish late Quaternary chronologies in various parts of the world, including central Alaska. Initially, some unusually young estimates were obtained for the Old Crow tephra from the Halfway House exposure in central Alaska. An apparent age of 86 ± 8 (1σ) ka for loess above and below the tephra (Wintle and Westgate, 1986), apparent ages of 99 ± 15 and 123 ± 20 ka for separate experiments on fine-silt-sized glass-rich fractions of tephra, and an apparent age of 108 ± 16 ka from loess just above the tephra at the Halfway House site (Berger, 1987) were obtained. Some of the loess samples of Wintle and Westgate were later analyzed by Berger, producing significantly older TL ages (Berger et al., 1994, fig. 17), in accord with the TL loess ages of Berger (1987). This illustrates the effect of inter-laboratory procedural differences. Loess immediately above the Sheep Creek tephra was determined to be 71 ± 3 ka (Wintle, 1987, written communication) and a glass age for the tephra of 80.9 ± 8.8 ka was reported (Berger, 1991).

This 80.9 ka TL result for the Sheep Creek tephra is discussed by Berger et al. (1996). Berger (1991) also reported an apparent age of 170 ± 27 (1σ) ka for the same glass-rich subfractions of the Old Crow tephra as analyzed by Berger (1987). The later experiments used a more effective procedure (preheating) to remove unstable TL (which produces age underestimates) than did Berger (1987).

In the mid 1980s, the TL sediment-dating method was applied in the central United States to dating loess that was beyond the range of radiocarbon dating. It was concluded by some geologists that the TL method severely underestimated the age of loess studied in Illinois and Iowa (Canfield, 1985; McKay, 1986; Norton and Bradford, 1985). In other parts of the world, TL was applied to loess of uncertain age, but presumed to be beyond the usual radiocarbon range of 40 ka. These early and later results have been reviewed in the context of a comparison of different TL sediment-dating procedures (Berger, 1988, 1989, 1994; Berger et al., 1994).

Because the TL method offers one of the few, or perhaps the only, direct means of dating loess, it was important to improve sample preparation, TL analyses procedures, and field collecting techniques. Vigorous work to achieve these objectives in this research stage of the method has been undertaken by Berger (e.g., 1985, 1987, 1990, 1991; Berger et al., 1991, 1992, 1994).

Consequently, using the improved TL methods of Berger et al. (1992, 1994), Berger and Péwé in 1991 undertook a project to apply TL dating to loess stratigraphically above and below the Eva Forest Bed. The objective was to determine more closely the numeric age of the major erosional unconformity that lies between the Goldstream Formation (Wisconsin) and the Gold Hill Loess (pre-Wisconsin; Fig. 4). Péwé began such a project in 1987 as a joint effort with Ann Wintle, then of the University of London. Péwé collected many TL samples from Wisconsin and pre-Wisconsin loess from several sections near Fairbanks, but only one TL date from the Goldstream Formation was produced (Péwé and Reger, 1989, fig. 7). In this context, loess samples were collected above and below the unconformity for this study (Fig. 24) by Berger and Péwé in 1991 and by Péwé in 1993. As an age check, Berger collected and dated loess just above and below the 140 ± 10 ka Old Crow tephra in a road cut on the south side of Birch Hill (Péwé, 1952a, figs. 21, 23; 1955, pl. 2, fig. 2, pl. 3, fig. 2; Fig. 3). The TL dates of 128 ± 22 ka and 144 ± 22 ka were obtained above and below the tephra respectively (Berger and Péwé, 1994; Berger et al., 1996).

Presented here are 14 preliminary TL dates from loess sections exhibiting the Eva erosional unconformity, the lower 3 m of the Goldstream Formation, and the upper 3 m of the Gold Hill Loess with enclosed tephra (Fig. 24). The sections are from low hill tops (Gold Hill, College Hill) and slopes (Fig. 3). The Eva Forest Bed is not present at the unconformity at most of these sections. At Dawson Cut, the unconformity has TL dates of 61 ± 6 ka above and 156 ± 29 ka below. Just below the unconformity at Sheep Creek Cut a TL date of 115 ± 19 ka was determined. At both of these localities TL dates indicate an age of the Sheep Creek tephra of >180 ka and <205 ka (Berger et al., 1996; revised from Berger and Péwé, 1994). From the three hilltop sections (Gold Hill, 2; College Hill, 1), four TL dates above the unconformity are from 66.7 ka to 107 ± 17 ka, and dates below the unconformity are 155 ± 33 ka and 207 ± 40 ka (Fig. 24; Berger, 1992, unpub. data). A TL date of 136 ± 20 ka was obtained on Gold Hill Loess 25 cm below the unconformity at upper Eva Creek (Figs. 14C, 24).

These new, preliminary TL loess dates constitute the strongest evidence yet available for the 125 ka age of the unconformity and of the associated Eva Forest Bed.

Summary of age discussion. Stratigraphic mapping of the associated loess deposits in glacial and periglacial areas of central Alaska suggested a last interglacial age for the Eva Forest Bed even before the days of widespread availability of absolute ages (Péwé, 1957, 1958, 1965b, 1965c). Indirect dating of the last interglaciation using climatic history based on the $^{18}O/^{16}O$ isotope ratio in fossil foraminifera tests in deep-sea piston cores (Emiliani, 1955) suggests a date of 127 ± 6 ka for the Oxygen-Isotope Stage 5/6 boundary, the beginning of the last interglaciation (Imbrie et al., 1984; Fig. 23). Martinson et al. (1987) also indicate that Substage 5e existed from 122 ± 2 ka to 125 ± 3 ka. However, based on uranium-series age measurements on raised coral terraces at some localities, the last interglacial period is believed to have begun somewhat earlier and lasted longer than indicated by the marine oxygen-isotope record: 122.5 to 125.1 ka (Bloom et al., 1974); 112 ± 6 ka to 137 ± 11 ka and 117 ± 3 ka to 141 ± 6 ka (Ku et al., 1974, and Muhs and Szabo, 1991) respectively; 120 ka to 132 ka (Chen et al., 1991); and 122 ± 8 ka to 152 ± 25 ka and 115 ± 10 ka to 160 ± 18 ka (Sherman et al., 1993). Sherman et al., (1993)

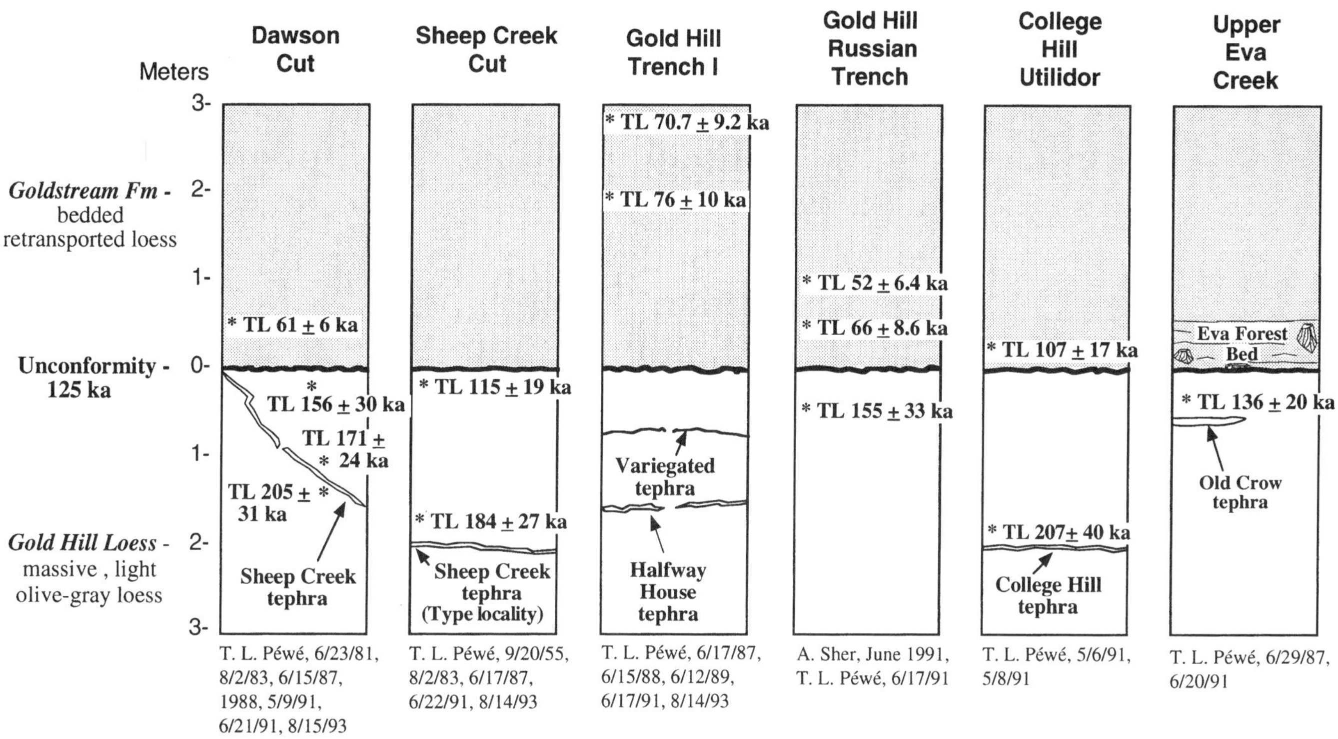

Figure 24. Stratigraphic sketches of preliminary thermoluminescence (TL) dates and tephra layers in Gold Hill Loess within 3 m above and below the Eva Interglaciation unconformity. The unconformity represents an erosional period during the Eva Interglaciation, 125 ka. Sections are from hill tops or valley slopes near Fairbanks, Alaska (Fig. 3), and are not from valley bottoms. Gold Hill Loess at Dawson Cut exposure is a tilted fault block under an angular unconformity. Eva Forest Bed is not present at all exposures. Stratigraphy by Péwé and TL sample collecting and dating by Berger (Berger and Péwé, 1994; Berger et al., 1996; and Berger, 1992–1995, unpub. data). Tephra identification by J. A. Westgate, B. Stemper, and S. J. Preece. College Hill tephra collected by Péwé in 1991. College Hill is site of University of Alaska (Fig. 3).

report further that these dates are from what are believed to be two high stands of the sea during the last interglaciation. Winograd et al., (1990, 1992) report that stable isotope analyses ($^{18}O/^{16}O$) made from vein calcite in Nevada suggests that the duration of Oxygen Isotope Substage 5e was from 120 ka to 140 ± 2 ka (Fig. 25).

Numeric dating of wood from the Eva Forest Bed and loess and tephra beds above and below the forest bed or associated erosional unconformity permits here a more refined estimate of the age and duration of the last interglaciation and the Eva Forest Bed in unglaciated east-central Alaska (Fig. 25). The unconformity at the top of the Gold Hill Loess (Fig. 4) is believed to mark the beginning of the interglaciation. Old Crow tephra (140 ± 10 ka) is near the top of the Gold Hill Loess, occurring actually at the unconformity (Figs. 10, 14a), or at various distances below the unconformity (1 to 5 m), depending upon how much of the Gold Hill Loess was eroded (Figs. 14B, 14C). At the Palisades (Begét et al., 1991) and at Birch Creek (Edwards and McDowell, 1991; Fig. 2) the Old Crow tephra is reported to be 5 m below a forest bed. At the Halfway House exposure, Old Crow tephra is 4 m below the Halfway House tephra (Preece, 1991). The Halfway House tephra is about 1.5 m below the unconformity at Gold Hill Trench 1 (Fig. 24) and about 50 cm below the unconformity at West Dawson exposure (Fig. 3; Péwé, field notes). Péwé collected the Halfway House tephra and at West Dawson it was identified by J. A. Westgate (1987, written communication).

If the youngest TL date of 115 ± 19 ka at the top of Gold Hill Loess at Sheep Creek Cut is used, the start of the Eva Interglaciation would be within the range 96 ka to 134 ka (Fig. 25). The estimate for the end of the interglaciation is suggested to be from 90 ka to 124 ka by using a TL date near the base of the Goldstream Formation in the College Hill Utilidor excavation (Figs. 24, 25).

We believe that despite the use of preliminary TL dates at this state of knowledge of the TL method, the plotting of the best age estimates now available indicates that the age of the Eva Forest Bed and the duration of the Eva Interglaciation can be safely generalized as about a few thousand years each side of 125 ka (Fig. 25).

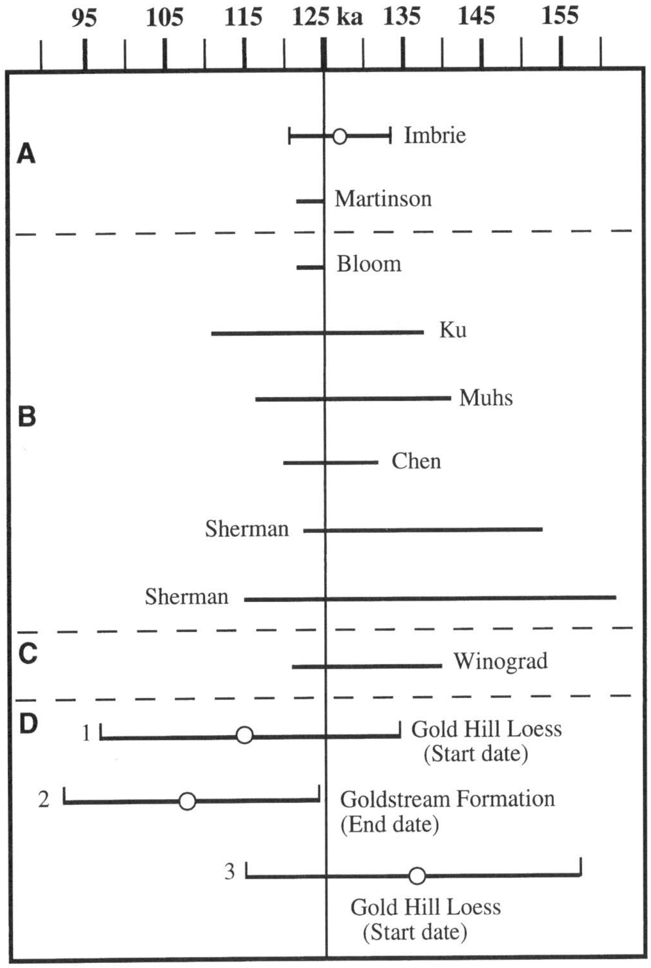

Figure 25. Plot of estimates of age and duration of last interglaciation (Oxygen-Isotope Substage 5e) based on (A) $^{18}O/^{16}O$ isotope ratio of fossil foraminifera (Imbrie, et al., 1984; length of bar is position of Oxygen-Isotope Stage 5/6 boundary; Martinson et a;/. 1987); (B) Uranium-series age measurements of raised coral terraces (Bloom et al., 1974; Ku et al., 1974; Muhs and Szabo, 1991; Chen et al., 1991; Sherman et al, 1993); (C) $^{18}O/^{16}O$ isotope ratio in vein calcite from U.S. Great Basin (Winograd et al., 1990); compared with age and duration of Eva Interglaciation based on (D): Thermoluminescence dates on Gold Hill Loess and Goldstream Formation near Fairbanks, Alaska (this report). Bar 1: TL date of Gold Hill Loess just under unconformity. Start date of interglaciation (Sheep Creek Cut, Fig. 24). Bar 2: TL date of loess at base of Goldstream Formation just above unconformity. End date of interglaciation (College Hill Utilidor, Fig. 24). Bar 3: TL date of Gold Hill Loess just under unconformity. Start of interglaciation (Eva Creek, Figs. 14C, 24).

PALEOENVIRONMENTAL INTERPRETATION

Preliminary statement

Reconstruction of the environment of the Eva Interglaciation is one of the most important aspects of the study of global warming as reflected in central Alaska. To demonstrate an interglacial environment we must show that a harsher glacial environment existed before and after the warmer forested stage in the periglacial area of east central Alaska. In addition to the taxa and distribution of trees in the interglacial forest of the Yukon-Tanana Upland and evidence for climate warming from physical evidence of permafrost thawing, we present here for the first time the dendrochronology of trees from the Eva Forest Bed (PMB) and comparisons of wood from the modern boreal forest and also comparisons of $^{13}C/^{12}C$ isotopic ratios of wood from Eva Forest Bed trees with modern wood from the Fairbanks area (SWL).

Environment of the loess formations bracketing the Eva Forest Bed

Early history of investigation. After the early conception of the existence of an interglacial event (the Eva Interglaciation) based on erosion and thawing of loess (Péwé, 1952a) and growth of a spruce-birch boreal forest, the fauna and flora of the late Gold Hill Loess and Goldstream Formation of retransported loess were investigated in a reconnaissance manner. Péwé (1950, 1952a, 1952b, 1955) demonstrated that pre– and post–Eva Forest Bed times were periods of great loess deposition, thought to coincide in time with major glacial advances in the Alaska Range 80 km to the south and elsewhere in the world (Capps 1932; Péwé, 1952b, 1954, 1955; Péwé et al., 1953). These times obviously had environments of more extensive permafrost and colder climates than during the Holocene and the Eva Interglaciations. Even though ice wedges of late Gold Hill Loess time are no longer present in central Alaska, ice wedges and permafrost of Wisconsin age still exist (Fig. 6) and illustrate the harsh periglacial environment of that time.

Examination of hundreds of frozen ground mining exposures in the late 1940s and early 1950s in central Alaska indicate an almost complete lack of specimens of frozen wood macrofossils in Wisconsin and pre-Wisconsin loess deposits (Péwé, 1952a; Repenning et al., 1964; Fig. 12). Apparently a treeless environment existed during these periods. A small collection of plant macrofossils was studied from retransported Wisconsin loess by Chaney and Mason (1936). Hansen (1953) collected a few Holocene samples for tree-pollen study. None of these early samples, however, was tied to the stratigraphy.

Initially, no palynological data were available to provide information of the harsher environments in central Alaska; therefore, a few specimens were collected in the late 1940s and early and middle 1950s (T. L. Péwé, field notes). Samples from a 1.5 m-thick peat-bed exposure 10-m long at the base of the Goldstream Formation at the north side of Ester Island (1949; Figs. 2, 22) and from the upper part of the Gold Hill Loess at West Dawson mining cut adjacent to Goldstream Valley (1955; Fig. 3) were examined by E. S. Barghoorn of Yale University. He reported that the 1949 material indicated a treeless tundra environment, colder than today, with abundant growth of *Equisetum*, grasses, sedges, herbs, and dwarf trees (birch and willow). Spruce pollen was totally lacking (E. S. Barghoorn, Nov. 23, 1954, written communication). The 1955 material revealed that the peat represented a locally treeless, swampy area some distance removed from alder and birch thickets (*Alnus* and *Betula* were conspicuously low). Although ferns are probably locally over-represented, a relatively sparse tree (spruce) cover is reasonable.

To compare this paleoenvironment with the nearby modern surface vegetation from the bottom of forested Goldstream Valley (Sec. 5, T.1.S., R.1W.; Fig. 3), two surface peat samples were collected (October 10, 1955; November 1, 1955) for pollen analyses (T. L. Péwé, 1955, field notes). These samples showed respectively a tree-pollen percentage of 53% and 33% (based on spruce and pine only, excluding alder and birch). These are conspicuously, and significantly, higher than tree-pollen percentages in the nearby Gold Hill Loess peat sample (E. S. Barghoorn, 1956, written communication). The high percentage of spruce pollen (probably white spruce) from the modern forested Goldstream Valley is similar to the 50% modern white spruce pollen obtained by Anderson and Brubaker (1986) for the center of the Yukon Flats on the north side of the Yukon Tanana Upland (Fig. 2). An early understanding of the absence or presence of wood macrofossils and these unpublished pioneer pollen investigations in the late 1940s and early and middle 1950s cited above, "provide the first botanical evidence that during Pleistocene intervals of harsher climate, dramatic lowering of treeline occurred, effectively eliminating trees from many areas of interior Alaska" (Ager and Brubaker, 1985, p. 363).

The record of abundant vertebrate fossils of large and small mammals in the Gold Hill Loess and Goldstream Formation, which bracket the Eva Interglaciation deposits, suggested a treeless "tundra" environment. Fossils of such mammals as ground squirrels, other tundra rodents, sheep, and antelope (*Siaga*) were known in the late 1940s and middle 1950s and many such fossils were collected in situ, mainly from the Goldstream Formation (Péwé, 1952a, 1957; Geist, 1953; Repenning et al., 1964).

Modern systematic studies. Systematic studies of flora and fauna of the Gold Hill Loess and the Goldstream Formation began in 1964 and 1965 when R. Dale Guthrie and John V. Matthews, Jr., of the University of Alaska started intensive examinations of the sediments. More than 60 tons of sediment were collected by thousands of bucketfuls from stratigraphic sections in mining exposures at Eva Creek, Sheep Creek Cut, Dawson Cut, and Ready Bullion Bench Cut (Fig. 3) and washed through 40 mesh screens at the sites. The residue was later examined in detail for Pleistocene pollen, beetle, and small mammal remains.

Pollen studies indicated that the silt was deposited in a cold, treeless environment (Matthews, 1968, 1970). In addition to those localities mentioned, tundra conditions were indicated in a silty peat >39,900 B.P. from the McGee Cut in the Tofty area (Figs. 2, 12; Matthews, 1968, 1970). An exception to the tundra environment was indicated by a study of pollen in a drill core from the Goldstream Formation in Isabella Basin near Fairbanks (Fig. 3). A possible mid-Wisconsin interstadial environment at >31,900 ka is recorded by higher spruce, birch, and alder pollen count (Matthews, 1974). The silt collected from the mining exposures was also studied for insect remains (beetles) by pioneer investigations of Matthews. His examination of the coleopterus fauna, mainly the ground beetles (carabid), also revealed that the environment was a cold treeless tundra (Matthews, 1968, 1970, 1974).

The most informative biological studies outlining the environment of the harsh, glacial, loess-depositional times are those dealing with the small and large late Pleistocene mammals. Detailed studies by Guthrie (1968a, 1968b, 1982, 1985) have extended over three decades and climaxed in his 1990 classic book entitled, *Frozen fauna of the Mammoth Steppe*. Guthrie (1990) demonstrated for the first time that this land north of the ice sheets in Alaska, Asia, and Europe was an extinct cold, arid, highly productive treeless grassland with a high carrying capacity. Called the Mammoth Steppe, it was not a boggy, unproductive tundra.

Small mammals such as the lemmings, ground squirrels, and voles thrived in the treeless well-drained soils (Guthrie, 1968a). They do not live in the lowlands of interior Alaska today. The large mammal population was dominated by grazers, such as mammoth, bison, and horse, and to a much lesser degree, by elk, sheep, and siaga antelope (Guthrie, 1968b). Bison, horse, and antelope require firm, dry terrain, not boggy country, and most of the large grazers of the Pleistocene on the Mammoth Steppe could not tolerate the snow drifts of today.

However, Elias et al., (1996) provide evidence from peats cored from beneath the seas that now cover the Bering land bridge that in late Pleistocene time this lowland terrain near the marine environment was not steppe but largely mesic shrub tundra. As Guthrie mentions (January 29, 1997, written communication), the pollen and insect data used by Elias are mostly from post-14 ka, basically postglacial, and represent a special regional area. These findings in no way negate the presence of a widespread cold grassy steppe stretching from Alaska across northern Asia and east to England during the latter part of the Pleistocene

In summary, physical and biological evidence demonstrate that a cold, harsh environment existed in late Gold Hill Loess time, just prior to the Eva Interglaciation, as well as in Wisconsin time, immediately after the interglaciation. This cold period was dominated by dust storms and loess and tephra deposition in an arid, treeless, grassy steppe. This environment has also been reported at the time of deposition of Old Crow tephra at the Porcupine River in Canada near the Alaska border (Schweger and Matthews, 1985). This falls in line with information from ice cores in both the Arctic and Antarctic that during the past 100 ka the atmosphere was extremely dusty during cold periods and much less dusty during warm periods (Thompson and Thompson-Mosley, 1981a, 1981b; Petit et al., 1981).

Snowfall was less than today and much of it was blown clear, permitting deep freezing. Permafrost with active ice wedges was widespread, in contrast to discontinuous permafrost and inactive ice wedges of today that are more or less restricted to lower slopes and creek-valley bottoms. The past active growth of ice wedges implies that the mean annual air temperature (MAAT) was –6 to –8 °C or colder (Péwé, 1966). July air temperatures in the southern part of the Yukon-Tanana Upland along the upper Tanana River (Fig. 2) for the Wisconsin period were calculated from present lapse rate and the position of snowline in Wisconsin time. Snowline was 600 m lower than today. The mean July air temperatures at Fairbanks and Northway, in the upper Tanana River Valley, for example, in Wisconsin time were 11.5 °C and 9.9 °C respectively, a drop of 4 °C and 4.8 °C from today (Péwé, 1975a, p. 108–109).

Dendrochronology of Eva Forest Bed trees and comparisons with modern trees

Introduction. Tree rings may record year-to-year fluctuations in environmental conditions such as soil moisture, temperature, atmospheric composition, disturbance regimes, or other factors depending upon the species and location under study (Fritts, 1976; Hughes et al., 1982; Schweingruber, 1987; Cook and Kariukstis, 1990). Characteristics of tree rings that have been intensively examined in modern trees include physical parameters such as ring width, ring density, and unusual ring structures such as scars or frost rings (e.g., Fritts and Swetnam, 1989), and chemical parameters such as $^{13}C/^{12}C$ ratios (e.g., Leavitt and Long, 1989; Leavitt, 1993).

There has been considerable dendroclimatic research on modern white spruce trees from the boreal forest in central Alaska. Much of this research indicates that present-day white spruce growth is strongly influenced by temperature regimes (Blasing and Fritts, 1975; Larsen, 1980; Cropper, 1982; D'Arrigo et al., 1992), although precipitation may also limit growth in more interior stands, especially on south-facing sites (L. A. Viereck, 1994, personal communication). Chronologies of ring width and ring density, particularly maximum latewood density, have been developed from modern Alaskan white spruce trees for purposes of climate reconstruction (Blasing and Fritts, 1975; Cropper, 1982; D'Arrigo et al., 1992; Schweingruber, 1983, unpub. data). To assess qualitatively the annual variability of climate during the time when the Eva Forest Bed trees grew, descriptive statistics of measurements of both ring width and density in annual rings from the Eva Forest Bed white spruce trees were compared to similar statistics from modern white spruce trees from central Alaska for which climate relationships are known.

Ring width and density in Eva Forest Bed and modern trees. White spruce wood samples from the Eva Formation are summarized in Table 1. For examination of tree-ring structure, all samples were first surfaced to 400 grit sandpaper using a belt sander and hand sanding. Ring numbers were counted on each sample and notes were made when unusual ring structures such as reaction wood or scars were observed within the ring series. Reaction wood is an area of expanded cell structure generally formed in response to a tree leaning off center, and characterized in conifer species by eccentric growth on the downhill portion of the ring circumference (Fritts, 1976). Ring widths were measured using a rotary encoder-measuring system connected to a microcomputer (Robinson and Evans, 1980). At least two radii were measured on most samples, although only one radius was measured in an area of the most uniform growth on samples with widespread areas of reaction wood or otherwise lobate growth patterns (Table 1, Fig. 26).

On selected samples, ring density was measured by X-ray densitometry (Schweingruber et al., 1978). Most of the white spruce samples from the Eva Forest Bed exhibited areas of reaction wood and these areas were unfortunately not suitable for density measurements owing to the expanded cell structures. Both methods used for measuring ring widths and densities are well-established techniques in tree-ring research and similar to methods used to derive the same parameters in the modern Alaskan white spruce samples used for comparison.

Five descriptive statistics were calculated and used to quantify and compare variation in width and density time series from the modern Alaskan and Eva Forest Bed samples. These statistics are often used to compare tree-ring series from different sites or species (c.f. Fritts, 1969; Fritts and Shatz, 1975; Fritts, 1976). The statistics may vary since the environmental and physiological processes responsible for formation of the tree-ring time series differ. These statistics are (1) average maximum measurement, (2) mean annual measurement, (3) mean standard deviation, (4) first-order autocorrelation, and (5) a statistic unique to dendrochronology, mean sensitivity (Douglass, 1936; Fritts, 1976). All of the statistics were calculated using individual ring measurements. The average maximum measurement is the maximum measurement from each individual tree radius averaged for all radii that make up a chronology. The mean annual measurement, standard deviation, first order autocorrelation, and mean sensitivity were first calculated for each radius and then averaged for the site. The average mean sensitivity of time series x (ms_x) is calculated as

$$ms_x = \frac{1}{n-1} \sum_{t=1}^{t=n-1} \left| 2(x_{t+1} - x_t)/(x_{t+1} + x_t) \right|$$

where x_t and x_{t+1} are yearly ring widths and n is the total number of years in the series. While mean sensitivity may range from 0 to 2, most ring-width series range from approximately 0.10 to 0.80. Program COFECHA (Holmes, 1983) was used to calculate statistics for each time series.

Two sets of modern Alaskan white spruce tree-ring data were compiled to compare to samples from the Eva Forest Bed. The first set was modified from Cropper and Fritts (1981) and Cropper (1982) and contains first-order autocorrelations, standard deviations, and mean sensitivities from only ring-width chronologies (Table 6). The second set was compiled from raw measurements from X-ray density scans from six white spruce sites archived in the International Tree-Ring Data Bank (ITRDB; Table 7). These six sites were collected and measured by Fritz Schweingruber of the Swiss Federal Institute of Forest Research. For this second set of modern data, statistics for ring width and density measurements were computed using program COFECHA (Holmes, 1983).

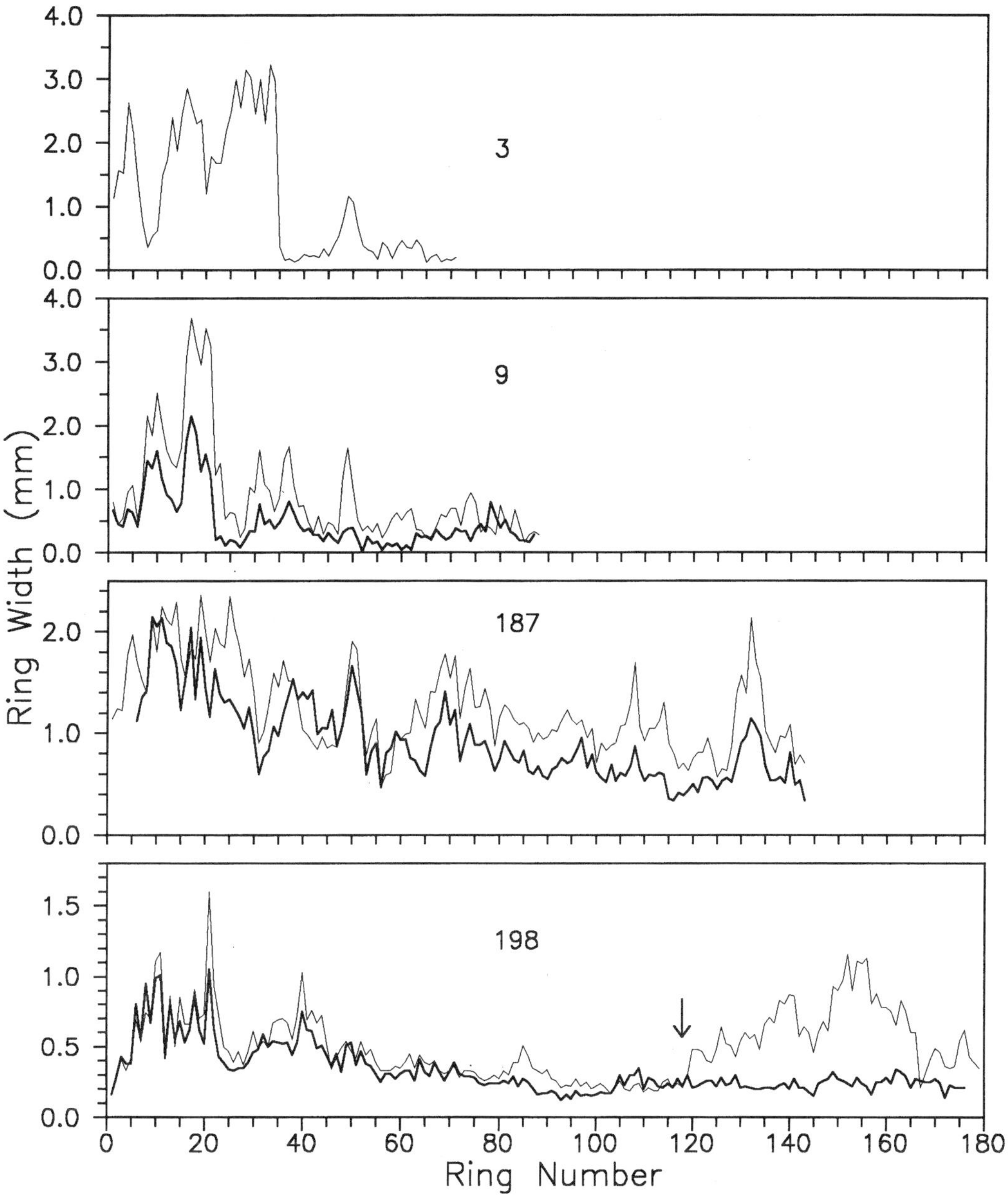

Figure 26. Plots of ring-width series from white spruce samples from the Eva Forest Bed (Table 1). For sample nos. 9, 187, and 198, two measured radii are plotted. The decrease in ring width on most radii is consistent with patterns seen in many species growing in relatively open canopy stands, where individual tree competition is minimized (see Fritts, 1976; Cook and Kariukstis, 1990). Increase in ring width due to areas of compression wood are evident on several of the radii, most notably on sample 198 beginning around ring number 118 on one radius (arrow in bottom plot).

General ring characteristics of the Eva Forest Bed samples. Annual ring boundaries were very distinct in the Eva Forest Bed samples (Figs. 16 and 17; cover photograph) with overall ring morphology indistinguishable from modern Alaskan white spruce trees. The Eva wood is sound and in generally excellent preservation. False rings (interannual latewood bands) are rare to nonexistent in modern white spruce and no false rings were seen in any of the trees from the Eva Forest Bed. In addition, no frost rings were found in any of the Eva Forest Bed samples. Frost rings are distinctive bands of disrupted cells formed when a hard frost occurs during the period of cell formation (Fritts, 1976).

As was noted in the preceding section, areas of reaction wood were seen on most of the Eva Forest Bed samples, suggesting that these trees were growing on unstable soils. Sample 187 exhibited an almost spiral pattern of reaction wood (Douglass, 1940; Telewski, 1988). This pattern of reaction wood is

TABLE 6. STATISTICS OF MODERN WHITE SPRUCE RING WIDTH MEASUREMENTS*

Site	First-Order Autocorrelation	Standard Deviation (mm)	Mean Sensitivity	Latitude	Longitude	Elevation (m)
Hope Creek	0.42	0.25	0.23	65°41'	145°58'	975
12 Mile Summit	0.59	0.25	0.22	65°21'	145°58'	914
Lower Goldstream	0.82	0.26	0.13	64°54'	147°51'	152
Glove Creek Bluff	0.40	0.34	0.28	64°13'	145°54'	274
Salcha Bluff	0.64	0.35	0.23	64°28'	147°02'	213
Fort Yukon-White Eye	0.64	0.28	0.19	66°32'	145°08'	152
Big Delta	0.59	0.23	0.16	63°53'	146°02'	732
Buchanan Creek	0.53	0.24	0.18	63°54'	146°56'	914
Newman-Dry Creek	0.48	0.22	0.18	64°06'	147°17'	914
Wood River	0.58	0.26	0.19	64°07'	147°37'	762
Home Creek	0.56	0.22	0.17	63°49'	148°26'	671
Stephens Village	0.63	0.25	0.16	66°00'	149°04'	
White Mountains	0.46	0.23	0.20	66°00'	147°00'	975
Alaska Range	0.49	0.18	0.15	63°20'	149°00'	914
Atlasta	0.55	0.20	0.14	62°06'	145°41'	305
Eureka	0.55	0.24	0.18	61°58'	146°53'	684
Talkeetna	0.61	0.32	0.20	62°17'	150°10'	107
Chulitna	0.62	0.22	0.16	62°56'	149°23'	884
Cantwell	0.88	0.36	0.15	63°23'	148°52'	884
Moody	0.85	0.39	0.19	63°45'	149°01'	884
Chitina	0.80	0.43	0.20	61°38'	144°31'	610
Paxson	0.37	0.17	0.15	63°05'	145°29'	610
Gakona	0.57	0.28	0.21	62°18'	145°09'	457
Slana	0.65	0.25	0.16	62°38'	144°26'	
Menasta	0.84	0.23	0.11	62°51'	143°56'	
Gold Creek	0.43	0.17	0.15	64°06'	140°49'	1,097
Hermanns Cabin	0.80	0.27	0.13	65°20'	147°30'	
Dawsons Junction	0.78	0.33	0.16	64°10'	141°30'	
Chandalar Lake	0.50	0.17	0.13	67°30'	148°30'	
Mount Fairplay	0.59	0.24	0.17	63°50'	142°00'	
Salcha River Headwaters	0.66	0.30	0.19	64°55'	144°00'	
Herron	0.93	0.31	0.09	63°05'	153°00'	305
Lake Sylva Summary	0.58	0.20	0.14	67°00'	148°00'	785
Yukon River	0.51	0.15	0.12	65°30'	150°00'	
Mean	0.61 ± 0.14	0.26 ± 0.06	0.17 ± 0.04			

*Modified from Cropper, 1982, and Cropper and Fritts, 1981.

very unusual in tree-ring series and is suggested to be the result of a gradual interannual rotational shift in the lean of a tree and resulting formation of reaction wood on differing areas of the tree's circumference (Telewski, 1988).

Statistics of ring width and density time series. Descriptive statistics of ring-width series measured from white spruce samples from the Eva Forest Bed are given in Table 8. Plots of ring-width measurements from four of these trees are shown in Figure 26. Ring widths of nine radii from five trees were measured for comparison to modern Alaskan white spruce samples. For comparison of ring densities, however, only four radii were measurable (Table 9), primarily owing to the presence of reaction wood in most of the Eva Forest Bed samples. Two of these four radii came from sample no. 187 with one each from sample nos. 9 and 198.

In comparing ring-width measurements from the Eva Forest Bed samples (Table 8) to the first set of modern sites (Table 6), it appears that the Eva samples had both more persistence from year to year (as measured by the first order autocorrelation) and more variability (as measured by the standard deviation and mean sensitivity) than most of the modern Alaskan white spruce chronologies. However, when comparing wood from the Eva Forest Bed with collections from the International Tree Ring Data Bank (Table 7), there is very little difference seen in either ring width or density-time-series statistics (Table 9). The only significant difference between modern Alaskan and Eva Forest Bed samples was more variability (as measured by standard deviation) in maximum latewood density in the Eva wood (Table 9). This difference in comparisons of the Eva Forest Bed trees with

TABLE 7. LOCATIONAL DATA FOR MODERN WHITE SPRUCE SITES FROM THE INTERNATIONAL TREE RING DATABANK

Site	Name	Latitude (N)	Longitude (W)	Elevation (m)
1	Denali National Park	63°40'	149°35'	750
2	Glenn Highway	61°20'	149°35'	100
3	Slana Bei Tok	62°50'	143°55'	600
4	Eureka Summit	61°50'	147°20'	960
5	Mt. Billy Mitchell	61°20'	145°15'	300
6	Northway Junction	62°50'	141°20'	600

TABLE 8. STATISTICS OF RING WIDTH MEASUREMENTS FROM WHITE SPRUCE SAMPLES, EVA FOREST BED

Sample*	Rings	Mean (mm)	Maximum (mm)	First Order Auto-correlation	Standard Deviation (mm)	Mean Sensitivity
3A	71	1.15	3.22	0.88	1.03	0.33
9A	88	0.93	3.68	0.85	0.81	0.38
9B	87	0.55	2.42	0.85	0.50	0.43
20A	23	1.27	3.14	0.84	0.85	0.27
187A	143	1.25	2.36	0.81	0.43	0.16
187B	143	0.93	2.14	0.85	0.41	0.18
187C	45	0.57	0.93	0.75	0.28	0.17
198A	179	0.51	1.60	0.80	0.25	0.18
198B	179	0.33	1.05	0.79	0.18	0.18
Mean				0.82 ± 0.04	0.53 ± 0.30	0.25 ± 0.10

*A, B, or C are different radii measured from the same sample.

the two sets of modern data may be due to the manner in which the descriptive statistical parameters were calculated. While it is unknown how the statistics in Table 6 were compiled, the methods used for the second set (Table 9) were exactly comparable to methods used to calculate the statistics from the Eva woods (using program COFECHA). Further discussion of the Eva Forest Bed trees will therefore concentrate on comparison with the second set of modern data (Table 9).

Variability of past climate. Physiological and environmental processes controlling tree growth may be highly variable between sites even for a given species. Local environmental conditions controlling tree growth (and hence formation of resulting time series) undoubtedly were responsible for some of the variation seen both among the modern Alaskan sites and between the modern sites and Eva Forest Bed samples. In general in Table 9, there is as much variability among the modern sites as between them and the Eva Forest Bed trees. Fritts (1976) lists several key microsite environmental factors that can lead to large differences in climate/growth relationships in trees and hence statistical characteristics of resulting tree-ring series. These factors include topographic position (e.g., canyon versus ridge top), aspect (e.g., north-facing versus south-facing slope), substrate, soil depth, elevation, and slope steepness. Most modern tree-ring chronologies collected for purposes of climate reconstruction (as was the purpose for most if not all of the sites listed in Tables 6 and 7) are from areas where large-scale climate variability would be the over-riding control on tree growth. For example, when looking for temperature responsive tree-ring series, trees growing in a cold-air drainage would not be sampled since growth may not show consistent response to overall temperature patterns for the general area. Many of the modern samples used for comparison to the Eva Forest Bed material are from upper elevation treeline sites, whereas the trees from the Eva Forest Bed were growing at elevations lower than 200 m. Other than elevation, other topographic and environmental conditions under which the Eva Forest Bed trees grew are not known. It is therefore impossible to account for microsite conditions between the different chronologies that may have been responsible for differences seen in ring-series statistics.

However, in general, and allowing for the very small sample size for the statistical parameters for wood from the Eva Forest Bed (four series), it is possible that environmental conditions responsible for formation of the tree-ring series during the Eva Interglacial period were comparable to conditions in operation today. Little difference was seen between modern Alaskan tree rings and the Eva Forest Bed ring series in any of the descriptive characteristics examined. The only significant difference between the Eva Forest Bed samples and the second set of modern Alaskan samples (Table 9) was that maximum latewood density appears to have been slightly more variable in the Eva rings. The fossil wood shows significantly greater standard deviation in maximum latewood density than any of the six modern sites. The mean sensitivity of maximum latewood density and mean sensitivity and standard deviation in latewood widths also tend to be higher in the Eva Forest Bed trees, although none of these three parameters was significantly different than the modern sites (Table 9). Generally, latewood width and maximum latewood density are correlated (e.g., Schweingruber et al., 1978), all of which suggests that there was more variability in latewood formation during the Eva Interglaciation than today. D'Arrigo et al. (1992) discuss the physiological and climatic controls on white spruce ring width and density, and they find that ring width and ring density, especially maximum latewood density, are often complementary in their response to differing seasons of climate conditions. Maximum latewood densities in modern Alaskan white spruce trees are most highly correlated with average yearly temperatures, whereas ring widths correlate inversely with spring temperatures. Greater variability in maximum latewood densities and possibly latewood widths in the Eva Forest Bed samples suggests that there may have been greater variability in annual temperatures during the period of time when these trees grew. However, whether the statistics in Table 9 reflect real differences in climatic conditions is certainly open to question. More series of both maximum latewood density and latewood width from Eva Forest Bed trees would be needed to confirm this observation.

TABLE 9. MEANS AND STANDARD ERRORS FOR DESCRIPTIVE STATISTICS OF FIVE DIFFERENT RING PARAMETERS FROM MODERN ALASKAN AND EVA FOREST BED WHITE SPRUCE SITES*

Site	Number	Maximum	Mean	Standard Deviation	First Order Autocorrelation	Mean Sensitivity
Ring Width		(mm)	(mm)	(mm)		
Denali National Park	13	1.45 ± 0.64^{a}	0.60 ± 0.44^{a}	0.22 ± 0.08^{a}	0.78 ± 0.61^{a}	0.18 ± 0.03^{a}
Glenn Highway	21	2.45 ± 0.75^{b}	0.91 ± 0.28^{b}	0.46 ± 0.17^{b}	0.82 ± 0.08^{a}	0.21 ± 0.02^{b}
Slana Bei Tok	18	1.76 ± 0.47^{a}	0.63 ± 0.20^{a}	0.31 ± 0.09^{c}	0.88 ± 0.07^{b}	0.17 ± 0.02^{a}
Eureka Summit	21	1.55 ± 0.60^{a}	0.69 ± 0.28^{a}	$0.28 \pm 0.09^{a,c}$	$0.83 \pm 0.11^{a,b}$	0.18 ± 0.03^{a}
Mt. Billy Mitchell	24	2.16 ± 0.61^{b}	1.05 ± 0.28^{b}	0.39 ± 0.14^{b}	0.79 ± 0.13^{a}	0.17 ± 0.02^{a}
Northway Junction	25	2.35 ± 0.91^{b}	$0.86 \pm 0.41^{a,b}$	0.48 ± 0.26^{b}	0.83 ± 0.08^{a}	0.25 ± 0.03^{c}
Eva Forest Bed	9	$2.28 \pm 0.96^{a,b}$	$0.83 \pm 0.35^{a,b}$	0.53 ± 0.30^{b}	0.82 ± 0.04^{a}	$0.25 \pm 0.10^{b,c}$
Earlywood Width		(mm)	(mm)	(mm)		
Denali National Park	13	1.34 ± 0.58^{a}	0.56 ± 0.38^{a}	0.22 ± 0.09^{a}	0.78 ± 0.14^{a}	0.22 ± 0.04^{a}
Glenn Highway	21	2.12 ± 0.61^{b}	0.77 ± 0.25^{b}	0.42 ± 0.15^{b}	0.81 ± 0.09^{a}	0.24 ± 0.03^{b}
Slana Bei Tok	18	1.45 ± 0.32^{a}	0.54 ± 0.17^{a}	0.28 ± 0.06^{a}	0.87 ± 0.06^{b}	0.19 ± 0.03^{c}
Eureka Summit	21	1.33 ± 0.52^{a}	0.58 ± 0.25^{a}	0.25 ± 0.08^{a}	$0.84 \pm 0.10^{a,b}$	0.19 ± 0.04^{c}
Mt. Billy Mitchell	24	1.93 ± 0.61^{b}	0.91 ± 0.25^{b}	$0.35 \pm 0.13^{a,b}$	0.79 ± 0.12^{a}	0.18 ± 0.02^{d}
Northway Junction	25	2.09 ± 0.85^{b}	$0.72 \pm 0.38^{a,b}$	0.43 ± 0.24^{b}	$0.83 \pm 0.09^{a,b}$	0.28 ± 0.04^{c}
Eva Forest Bed	4	$1.53 \pm 0.48^{a,b}$	0.78 ± 0.34^{b}	$0.25 \pm 0.11^{a,b}$	0.71 ± 0.08^{a}	0.19 ± 0.02^{c}
Latewood Width		(mm)	(mm)	(mm)		
Denali National Park	13	$0.40 \pm 0.18^{a,b}$	0.12 ± 0.05^{a}	$0.06 \pm 0.03^{a,b}$	$0.64 \pm 0.18^{a,b}$	0.21 ± 0.03^{a}
Glenn Highway	21	0.57 ± 0.39b	0.16 ± 0.04^{b}	0.08 ± 0.05^{b}	0.59 ± 0.13^{a}	0.24 ± 0.04^{b}
Slana Bei Tok	18	0.51 ± 0.30^{b}	0.11 ± 0.03^{a}	$0.07 \pm 0.04^{a,b}$	0.71 ± 0.11^{b}	0.23 ± 0.05^{b}
Eureka Summit	21	0.37 ± 0.18^{c}	0.11 ± 0.03^{a}	0.05 ± 0.02^{a}	0.54 ± 0.17^{a}	0.25 ± 0.04^{b}
Mt. Billy Mitchell	24	0.38 ± 0.13^{a}	$0.14 \pm 0.04^{a,b}$	0.05 ± 0.02^{a}	0.58 ± 0.18^{a}	0.23 ± 0.03^{b}
Northway Junction	25	0.42 ± 0.22^{a}	0.12 ± 0.03^{a}	$0.06 \pm 0.02^{a,b}$	$0.71 \pm 0.16^{a,b}$	$0.22 \pm 0.04^{a,b}$
Eva Forest Bed	4	0.63 ± 0.27^{b}	0.19 ± 0.04^{b}	0.11 ± 0.05^{b}	$0.61 \pm 0.16^{a,b}$	0.28 ± 0.05^{b}
Maximum Latewood Density		(g/cm^3)	(g/cm^3)	(g/cm^3)		
Denali National Park	13	0.78 ± 0.05^{a}	0.61 ± 0.06^{a}	0.07 ± 0.02^{a}	0.41 ± 0.22^{a}	0.09 ± 0.01^{a}
Glenn Highway	21	0.93 ± 0.04^{b}	0.78 ± 0.03^{b}	0.07 ± 0.01^{a}	$0.52 \pm 0.20^{a,b}$	0.06 ± 0.01^{b}
Slana Bei Tok	18	0.94 ± 0.04^{b}	0.76 ± 0.06^{b}	0.08 ± 0.01^{b}	0.66 ± 0.12^{b}	0.07 ± 0.01^{b}
Eureka Summit	21	0.86 ± 0.05^{c}	0.69 ± 0.06^{c}	0.07 ± 0.00^{a}	0.38 ± 0.17^{a}	0.09 ± 0.02^{a}
Mt. Billy Mitchell	24	0.94 ± 0.05^{b}	0.80 ± 0.04^{b}	0.06 ± 0.01^{a}	0.37 ± 0.13^{a}	0.07 ± 0.01^{b}
Northway Junction	25	0.93 ± 0.05^{b}	0.76 ± 0.05^{b}	$0.07 \pm 0.01^{a,b}$	0.61 ± 0.17^{b}	0.06 ± 0.00^{b}
Eva Forest Bed	4	0.93 ± 0.11^{b}	0.72 ± 0.07^{c}	**0.10 ± 0.02^{c}**	$0.55 \pm 0.25^{a,b}$	0.10 ± 0.2^{a}
Minimum Earlywood Density		(g/cm^3)	(g/cm^3)	(g/cm^3)		
Denali National Park	13	0.41 ± 0.08^{a}	0.30 ± 0.04^{a}	0.03 ± 0.01^{a}	0.68 ± 0.14^{a}	0.05 ± 0.01^{a}
Glenn Highway	21	0.43 ± 0.06^{a}	0.32 ± 0.01^{b}	0.03 ± 0.01^{a}	0.66 ± 0.15^{a}	0.05 ± 0.01^{a}
Slana Bei Tok	18	0.42 ± 0.08^{a}	0.31 ± 0.03^{a}	0.03 ± 0.01^{a}	0.64 ± 0.13^{a}	0.06 ± 0.01^{b}
Eureka Summit	21	0.42 ± 0.08^{a}	0.31 ± 0.03^{a}	0.03 ± 0.01^{a}	0.69 ± 0.13^{a}	0.05 ± 0.01^{a}
Mt. Billy Mitchell	24	0.38 ± 0.05^{b}	0.30 ± 0.02^{a}	0.02 ± 0.01^{b}	$0.62 \pm 0.13^{a,b}$	0.05 ± 0.01^{a}
Northway Junction	25	0.44 ± 0.06^{a}	0.30 ± 0.02^{a}	0.04 ± 0.01^{c}	0.56 ± 0.13^{b}	0.08 ± 0.01^{c}
Eva Forest Bed	4	$0.43 \pm 0.12^{a,b}$	0.33 ± 0.04^{b}	$0.04 \pm 0.03^{a,c}$	$0.63 \pm 0.34^{a,b}$	0.05 ± 0.01^{a}

*Modern site locations are given in Table 7. Significant differences between means for each statistic were tested using a nonparametric Kruskal-Wallis test. Differences between means are denoted by superscript letters (SAS Institute, 1986, $P < 0.05$). For example, all of the numbers with "a" superscripts have means that are similar in value, but significantly different than numbers with "b" or "c" superscripts. The only significant difference between the Eva Forest Bed statistics and those from modern sites is between standard deviations in maximum latewood density (highlighted in bold).

$^{13}C/^{12}C$ isotopic ratios of Eva Forest Bed trees and comparison with modern and Holocene trees

Tree-ring and wood $\delta^{13}C$ values, $=[(^{13}C/^{12}C_{sample}/^{13}C/^{12}C_{standard}) - 1] \times 1{,}000$ in ‰, have been used to infer past atmospheric chemistry (e.g., Stuiver, 1978; Leavitt and Long, 1988) and paleoclimate (e.g., Leavitt and Long, 1989; Libby et al., 1976). The availability of wood from Alaska spanning 125,000 years permits determination of plant $\delta^{13}C$ through time holding geographic location constant. Variations of $\delta^{13}C$ may reflect environmental differences between the prior (Sangamon) and current (Holocene) interglaciations. For example, temperature, humidity, cloudiness, and storminess have likely varied over the past 125,000 years.

Eleven samples from the Eva Forest Bed (9 spruce and 2 angiosperm trees), 15 samples from the Holocene between 10,000 B.P. and 1,000 B.P. (6 spruce and 9 angiosperms), and 9 modern trees (4 spruce and 5 angiosperms) were analyzed isotopically. The Holocene and modern wood was collected at sites near the Eva Forest Bed exposures (Tables 1, 2, and 3), and angiosperm trees of all ages represent *Populus*, *Salix*, and *Betula* species. A subsample representative of each piece of wood was obtained by milling thin wood cross sections to 20 mesh. Obtaining such samples of complete growth rings from each of many trees is important because of both inter-tree and intra-tree $\delta^{13}C$ variability of tree rings (Leavitt and Long, 1986).

The holocellulose component of the wood was isolated by the acid chlorite extraction method (Leavitt and Danzer, 1993) after initial removal of resin and oil extractives with toluene-ethanol. The holocellulose was combusted to CO_2 in the presence of excess oxygen. Purified CO_2 was analyzed on a VG Micromass 602C isotope ratio mass spectrometer, and $\delta^{13}C$ results calculated with respect to the international PDB (peedee belemnite) standard. Repeated combustion and analysis of a cellulose standard in conjunction with the Alaska tree isotopic analyses gave a precision of ±0.25‰ (±1 standard deviation, n = 11).

$\delta^{13}C$ *comparison of modern, Holocene, and Eva Forest Bed trees.* The $\delta^{13}C$ results are presented in Figure 27. They show the tendency for the modern angiosperm trees to be ^{13}C-depleted (more negative $\delta^{13}C$) with respect to the modern spruce trees. Leavitt and Newberry (1992) attempted to quantify differences between gymnosperm and angiosperm trees and concluded the

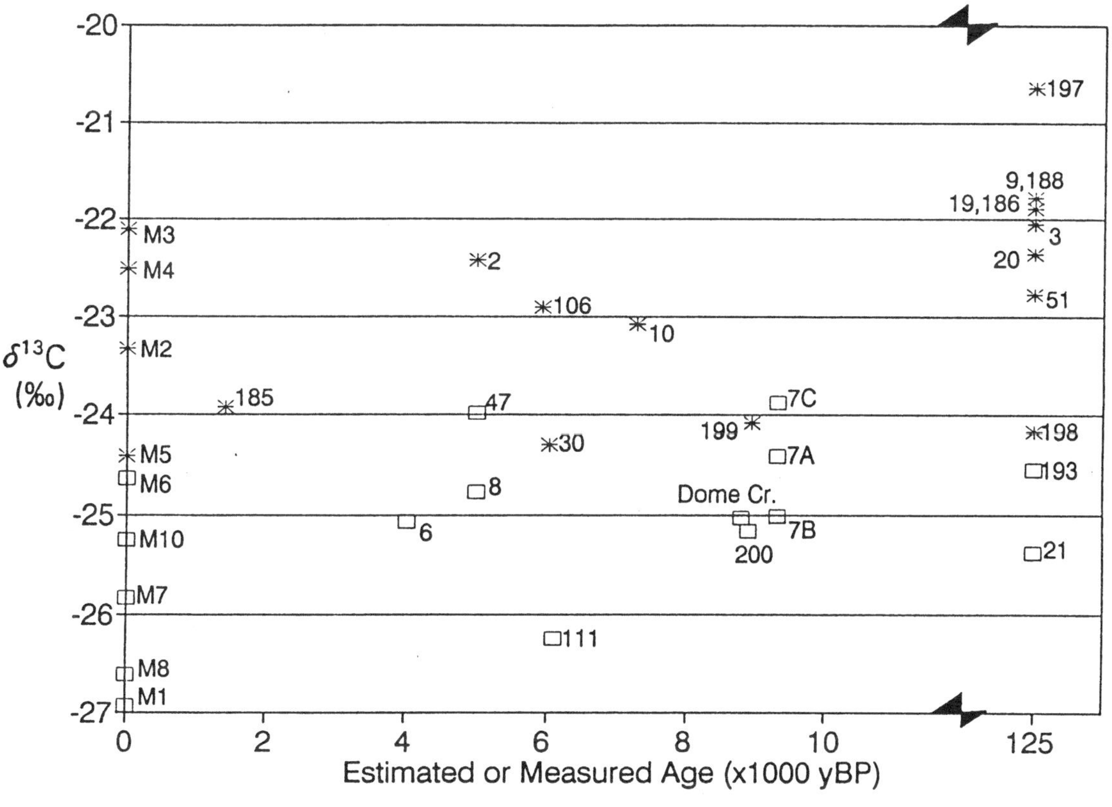

Figure 27. The $\delta^{13}C$ measured in the cellulose component of modern and fossil wood samples from the Fairbanks area, Alaska. The Holocene and Sangamon samples are from perennially frozen loess and retransported loess. Asterisk indicates spruce specimens; rectangle indicates angiosperm specimens (*Populus*, *Salix*, and *Betula*). The numbers refer to samples in Table 1 (125 ka Eva Forest Bed); Table 3 (Holocene-Giddings Forest Bed); and Table 2 (modern trees). All samples analyzed by S. Leavitt and all samples collected by T. L. Péwé, except sample no. 51 which was collected by M. E. Edwards at Birch Creek site and is not in Table 1.

angiosperms are typically ~1 to 3‰ isotopically more ^{13}C-depleted than the gymnosperms. The average gymnosperm-angiosperm $\delta^{13}C$ differences in the Alaska trees for modern, Holocene and Eva are 2.8, 1.4, and 2.9‰, respectively, falling within the previously established range.

The Holocene conifers average 1.4‰ more negative than those of the Eva Forest Bed period, whereas the modern angiosperm trees are 1.0‰ more negative than those of the Eva Forest Bed. Other studies of $\delta^{13}C$ variations in C_3 plant matter over the past 40,000 years have shown a similar magnitude of isotopic variation over the last deglaciation. Krishnamurthy and Epstein (1990) analyzed wood from a number of sites worldwide that had been previously ^{14}C dated. They found a $\delta^{13}C$ decrease of ~3 to 4‰ from ca. 18,000 B.P. to the Holocene, and believed the pattern of changes matched that of CO_2 changes measured in ice cores. They interpreted the $\delta^{13}C$ changes to represent $\delta^{13}C_{air}$, although subsequent ice core (Leuenberger et al., 1992) and C_4 plant (*Atriplex*) reconstructions (Marino et al., 1992) estimated a much smaller global $\delta^{13}C_{air}$ shift (~0.6‰), but in the opposite direction. Tabulating $\delta^{13}C$ values of wood and other C_3 plant matter over the past ca. 40,000 years, Leavitt and Danzer (1991, 1992) found evidence for a $\delta^{13}C_{plant}$ decrease of ~1‰ from ca. 18,000 B.P. to Holocene. This was interpreted as the effect of rising atmospheric CO_2 concentrations (~190 ppm at 18,000 B.P. and 270 to 280 ppm in the Holocene) on plant carbon isotope fractionation. Leavitt and Danzer (1992) also looked at a series of wood specimens from buried forests which spanned the period from 12,000 B.P. to the present in the Great Lakes region. However, after accounting for the isotopic differences between angiosperm and conifer trees, and for the more negative $\delta^{13}C$ values of modern trees because of inputs of recent ^{13}C-depleted CO_2 to the atmosphere, they found no net change in the $\delta^{13}C$ over that period of time. Van de Water et al. (1994) measured $\delta^{13}C$ in *Pinus flexilis* needles from packrat middens over the past 40,000 years and found a $\delta^{13}C$ decline of ~1.5‰ from ca. 18,000 B.P. to Holocene. They interpreted this as related to the effects of CO_2 concentration on water-use efficiency and plant carbon isotope fractionation.

The results presented in Figure 27 represent two interglacial periods rather than deglaciation, yet the Holocene spruce $\delta^{13}C$ is more negative than that of Eva Forest Bed time. This suggests environmental differences between the two interglacials. However, the picture becomes cloudy when angiosperms and modern spruce are considered. For example, there is no difference between average $\delta^{13}C$ values of the Eva and mid-Holocene angiosperm trees, suggesting no environmental differences. Additionally, the modern spruce average 0.4‰ greater than the Holocene spruce, when one would expect the modern to be significantly more negative due to atmospheric $^{12}CO_2$ enrichment over the past 200 years from burning fossil fuels (Leavitt and Long, 1988). If the average modern spruce and angiosperm values are "corrected" upward ~1.2‰ for the effect of fossil-fuel combustion since the beginning of the Industrial Revolution, their mean values then come in line with those of the Eva Forest Bed $\delta^{13}C$ averages. This suggests that after accounting for industrial effects, the Eva Forest Bed environment may be identical to modern. No such "industrial" correction can be applied to the mid-Holocene wood, so one is left with the dilemma of mid-Holocene spruce 1.4‰ more negative than the Eva spruce, but mid-Holocene angiosperm tree $\delta^{13}C$ identical to that of the Eva Forest Bed angiosperms.

Environmental implications of $\boldsymbol{\delta^{13}C}$**.** Full interpretation of the isotopic results requires consideration of the plant stable-carbon isotope fractionation model of Farquhar et al. (1982):

$$\delta^{13}C_{plant} = \delta^{13}C_{air} - a - (b - a)C_i/C_a,$$

where a and b are constants related to kinetic fractionation by ribulose bisphosphate carboxylase and during diffusion into leaves, respectively. There are currently no estimates of $\delta^{13}C_{air}$ from the last interglacial (ca. 125,000 B.P.). The C_i/C_a term is the ratio of CO_2 concentration inside the plant (leaf) versus that outside in the atmosphere. Evidence from ice cores suggests the Holocene and Sangamon have similar atmospheric CO_2 concentrations (C_a) in the range of 270 to 290 ppm and similar temperatures (Barnola et al., 1987; Jouzel et al., 1993). The C_i/C_a ratio, however, can be affected by not only variation in the atmospheric CO_2 concentration but by environmental factors which influence C_i such as light, humidity, nutrient status, soil moisture availability, and others (Francey and Farquhar, 1982). Thus, plant $\delta^{13}C$ is determined by many other factors in addition to $\delta^{13}C_{air}$.

Allowing for the relatively small number of samples analyzed, the general $\delta^{13}C$ similarity of both sets of interglacial wood samples suggests that neither $\delta^{13}C_{air}$ nor C_i/C_a values were significantly different between these two time periods. However, until $\delta^{13}C_{air}$ during Eva Forest Bed time is actually determined by some independent means (e.g., ice core measurements), we cannot exclude the possibility that Eva $\delta^{13}C_{air}$ could have been different than that of the Holocene and that the difference was compensated by dissimilar C_i/C_a ratios so that the $\delta^{13}C_{plant}$ values were approximately the same. Given the current ice core evidence of similar CO_2 concentrations during the two periods, such a scenario would require that some other environmental factor cause differences in C_i/C_a.

Summary of the environment of the Eva interglaciation Forest Bed

Introductory statement. We will summarize and interpret in a historical format the interglaciation environment of the Eva Forest Bed of the Yukon-Tanana Upland by reviewing several sets of data: (1) analyses of the taxa and distribution of trees; (2) dendrochronology and (3) isotope examination of the wood; (4) pollen data; (5) physical reaction of the landscape to climatic change, such as permafrost thawing and refreezing and rapid erosion, gullying, and block slumping of loess; in addition, (6) pollen, insects, and physical data from adjacent unglaciated northern Yukon Territory, Canada (Fig. 1); and (7) astronomical climatic inferences.

Botanical evidence—Yukon-Tanana Upland. From the early 1950s (Péwé, 1952a, 1965c, 1975a) the composition of the Eva Forest Bed has been considered to represent the typical taiga, or northern boreal forest, of central Alaska. Specimens from the frozen bed indicate a spruce-birch forest with at least the following taxa: white spruce (*Picea glauca*), black spruce (*Picea mariana*), paper birch (*Betula papyrifera*), poplar (*Populus* sp. probably cottonwood and aspen), alder (*Alnus* sp.), willow (*Salix* sp.), and *Sphagnum* moss. The size and frequency of spruce and birch are similar to representatives in today's taiga. Past widespread forest fires were similar to that of the modern boreal forest. Also, the ubiquitous spruce bark beetle was present then as now in the boreal forest of Alaska. Such a forest exists today (Fig. 7) from the south flank of the Brooks Range to the Anchorage area (Fig. 1), an area with a range of mean annual air temperatures from about –8 °C in the north to +2 °C in the south (Watson, 1959).

Dendrochronological studies on trees from the Eva Forest Bed and comparison with wood from the modern boreal forest (Table 9), suggest that environmental conditions were comparable to conditions today in central Alaska. Comparisons of $^{13}C/^{12}C$ isotopic ratios of wood from Eva Forest Bed trees with Holocene and modern wood from the Fairbanks area, suggest that neither $\delta^{13}C$ nor C_i/C_a values were significantly different between these two time periods. Unfortunately, no comparative $^{13}C/^{12}C$ isotopic ratio information of tree wood is available from central Alaska of harsh glacial time. Finally, at the Birch Creek site near Circle (Fig. 2), Edwards and McDowell (1989, 1991) report the presence in Eva Interglaciation silt of pollen of the cat tail, *Typha latifolia*, which is near its northern limit in the area today; this suggests climatic conditions at least as warm as present.

From only the botanical studies listed above we deduce the existence of (1) a forested interglacial environment and not an interstadial, and (2) an environment at least comparable to the lower elevations of the Yukon-Tanana Upland today. Discontinuous permafrost likely existed and inactive ice wedges were widespread. Such frozen ground conditions indicate a mean annual air temperature near –3 °C or –4 °C, as occurs today on the flood plain at Fairbanks and a mean July air temperature of about 15 °C.

Physical evidence—Yukon-Tanana Upland. Since the late 1940s, it has been known from geologic studies of the frozen loess, especially stratigraphic studies, that this interglacial interval in the Yukon-Tanana Upland was characterized by great and rapid thawing of permafrost and erosion of loess (Péwé, 1952a, 1957, 1965c, 1975a, 1989, 1992, in press). At the beginning of the Eva Interglaciation, with retreat of glaciers in the Alaska Range, major streams cut down (Blackwell, 1965; Péwé, 1965b; Péwé and Reger, 1983) permitting tributary streams to cut into creek-valley fills of frozen loess, especially on the south side of the Yukon-Tanana Upland (Fig. 2). Precipitation likely increased during the interglaciation, falling on the steppe vegetation-covered loess slopes that still existed in the early interglaciation, prior to the growth of the Eva Forest. Perhaps even more important, thawing of permafrost, and especially melting of massive ice wedges and pingo ice, provided extra water to slice into the easily erodible Gold Hill Loess. In this way much of the Gold Hill Loess was removed (see Fig. 4).

An increase, perhaps a rapid increase, of running water in a poorly vegetated loess environment and rapid permafrost thawing, are ideal conditions to erode the landscape in a spectacular fashion. Deep, vertical-sided gullies and steep-headed valleys formed, as well as sinkholes and natural bridges in loess—a typical thermokarst terrain. Such are known from the extensive loess deposits of China and from the modern active placer mining areas of central Alaska.

Massive, fine-grained loess is susceptible to slumping in large multiple blocks. Such gullying and slumping on upper and middle hillslopes apparently took place during the Eva Interglaciation, especially in the early part of the warm interval. Now rounded, parallel gullies and ridges in loess (see Péwé, 1955, pl. 3, figs. 1, 2, for aerial view, 1992) are relicts of this process. On lower slopes and in some valley bottoms, buried under Wisconsin and Holocene loess deposits, are large, multiple slump blocks of loess strikingly outlined by the tilting of white tephra beds in the massive silt (see Figs. 4, 8, 9, 11a, 12, 13, 22, 24).

After most of the angular sculpturing and slumping of the frozen loess, the micro-loess topography was smoothed and the Eva Forest became extensive. Thawing of permafrost continued until most, if not all, of the ground was thawed. The absence today of ice wedges, buried pingos, and any complete or partial mammal carcasses in the refrozen loess of pre-Wisconsin age provide evidence of deep thawing. Such features are known today only from frozen Wisconsin and Holocene sediments in the Yukon-Tanana Upland. Ice wedge casts of pre-Wisconsin age are rare in the loess because of the great amount of erosion on the hill slopes. They have been reported in the Yukon-Tanana Upland at Halfway House (T. L. Péwé, field notes), University of Alaska campus (Péwé, 1965c), and Shaw Creek Flats (Péwé, 1965b, fig. 5-22), but none of these is in loess.

The unique, and most informative, evidence for widespread, deep thawing of permafrost in this loess is the presence of perennially frozen, green Gold Hill Loess (frontispiece). This loess, with a bright green to olive green color, occurs in all fresh exposures of massive Gold Hill Loess underlying either the Eva Forest Bed and the Goldstream Formation, or just the Goldstream Formation. Almost all stratigraphic sketches in this report illustrate a green to greenish Gold Hill Loess (Figs. 8, 9, 10, 11, 12, 13, 14C, 22) in the freshly opened frozen exposures. Detailed studies in the 1940s and early 1950s of the stratigraphy, mineralogy, geochemistry, granulometric nature, and photography of the green and tan loesses (Péwé, 1952a), indicate that the iron coating on the minerals (mainly quartz) in the loess was reduced from ferric (tan) to ferrous (greenish) by deeply penetrating percolating surface water as the loess was thawed under a vegetative cover in anaerobic conditions. The thickest sections of green loess are in creek valley bottoms and they become thinner upslope. The thickness of frozen loess with green color is ~10 to 20 m. Refreezing in Wisconsin time has preserved the green color until exposed and thawed today. The thawing and percolation of downward-moving water probably occurred over a long time in an environment warmer than at present. The mean annual air temperature must have been warmer than 0 °C for the ice to melt and permafrost to thaw. The exact mean annual air temperature is not known, perhaps +1 or 2 °C.

Based on botanical, and now the physical data, from only the Yukon-Tanana Upland, we can now briefly summarize in more detail the environment of the Eva Interglaciation. The environment was warmer than the present interglaciation with a mean annual air temperature greater than 0 °C, perhaps +1 or +2 °C or warmer for the ice to melt and permafrost to thaw from the surface downward. The mean July air temperature was probably greater than the present 15.5 °C (Fairbanks) and precipitation was likely greater than that of the arid, Mammoth Steppe of glacial times. We think that the warming must have been early and rapid and must have resulted in much erosion and deep thawing in the loess, with the subsequent growth of a widespread boreal forest. Evidence of buried spruce (*Picea*) macrofossils indicates that fingers of the interior forest in Eva Interglaciation time may have penetrated northward through passes in the Brooks Range to be present in the valleys on the north side of the mountains (Hopkins, 1972; Péwé 1975a; Carter and Ager, 1989).

Adjacent Yukon Territory, Canada. About 50 km east of the Alaska-Canada border on the southside of the Porcupine River (Fig. 1) occurs a remarkable river-bluff exposure of frozen late Cenozoic sediments. Ch'ijee's Bluff lies just north of the Arctic Circle in the unglaciated Alaska-Yukon interior, in the far northern part of the boreal forest (Fig. 6). The local mean annual air temperature is about –8 °C to –10 °C. Quaternary studies in the area began in the 1960s (Schweger, 1989). In the 1980s careful, detailed, multidisciplinary work was undertaken on this and other bluffs by Matthews and Schweger. Thus, a valuable and clear interpretation of the local environment for about the past 200 ka or more is now known (Schweger and Matthews, 1985; Matthews et al. 1987, 1989, 1990). A summary of their interpretation of the nature of this local environment is considered here because of the strong multidisciplinary evidence of a warm interglaciation that strongly supports the conditions outlined for the Eva Interglaciation environment for the adjacent interior Alaska.

Matthews et al. (1990) confirm that the Old Crow tephra was deposited at a time much colder than now, and before the last interglaciation. Above the Old Crow tephra in the section are several lines of evidence for an interval having summers warmer than present, which they informally term the "Koy-Yukon interglaciation." Matthews et al. (1989) state: "The fluvial and lacustrine deposits contain plant macroremains of species like *Carex sychnocephala* and *Chenopodium gigantospermum*, neither of which extend as far north today. A similar pattern is exhibited by some of the insect taxa (such as *Bradycellus lecontei* and *Bembidion quadrimaculatum*), and the presence of traces of *Corylus* and *Typha* in the pollen spectra." Furthermore, extremely high frequencies of spruce are shown by the pollen work of Lichti-Federovich (1973). In addition, a series of large ice-wedge pseudomorphs are present, which may reflect regional thawing of permafrost (Westgate et al., 1983, figs. 3, 5).

Astronomical climatic inferences. Bartlein and Prentice (1989) indicate that during the last interglaciation, the July insolation anomaly at 65°N reached values almost 50% higher than 10,000 years ago. This suggests that the last interglaciation was much warmer and had a more rapid onset than did the Holocene. Watts (1988, p. 186) demonstrates that the last interglaciation in Europe also was much warmer than the Holocene and had a rapid onset. Heusser and King (1988, p. 220) likewise document a warmer last interglaciation in North America. Such data are consistent with the suggested rapid thawing of permafrost, gullying of loess, and deep or complete degradation of permafrost in the Yukon-Tanana Upland during the Eva Interglaciation.

CONCLUDING REMARKS

One hundred forty thousand years ago, a cloud of volcanic ash from the Aleutian Chain 1,000 km to the southwest passed over the Yukon-Tanana Upland, depositing a white gritty tephra layer on a cold, arid, treeless steppe, grass-dominated landscape (Westgate, 1988; Guthrie, 1990; Schweger and Matthews, 1985). This ash, now termed the Old Crow tephra, blanketed the loess-covered terrain and was soon buried by additional Gold Hill Loess and younger tephra layers. Permafrost was widespread, and in winter the ground cracked to allow the massive, ubiquitous ice wedges to receive moisture and continue to enlarge.

About 10,000 or 15,000 years later the windy, dry, harsh glacial climate, with a mean annual air temperature of perhaps –8 °C to –10 °C or colder in the lower slopes and valley bottoms of the Yukon-Tanana Upland, began to warm and precipitation increased. Evidently, this loess landscape at the end of glacial time exhibited exposed soil within an open grassy steppe (Guthrie, 1990). Before a protective boreal forest could be well established, global climatic change caused warmer temperatures that initiated thawing of the underlying permafrost and melting of the massive ice wedges. This extra running water, plus increased precipitation after the arid glacial time, cut into the treeless landscape of massive silt. Intensive gullying and slumping of loess blocks occurred first, and then smoothing of the thermokarst terrain occurred. With the declining of loess deposition, the warming of mean annual temperature to more than 0 °C, and the increasing of precipitation, the Eva Interglaciation boreal forest became widespread, and some plants (and insect taxa in Canada) extended even farther north in central Alaska than today.

During the Eva Interglaciation, the lower slopes and valley bottoms of the Yukon-Tanana Upland did not resemble a permafrost environment such as exists today. Forested open-system pingos (Péwé, 1982, fig. 61) were absent; as were large-scale ice-wedge polygon patterns, beaded drainage, and cave-in lakes, all of which exist in central Alaska today (Péwé, 1982, figs. 24, 60). The Eva boreal forest of central Alaska blanketed a terrain with a climate milder than now with a mean annual air temperature warmer than 0 °C, resulting in extensive degradation of permafrost.

REFERENCES CITED

Ager, T. A., and Brubaker, L. R., 1985, Quaternary palynology and vegetational history of Alaska, *in* Bryan, T. V., Jr., and Holloway, R., eds., Pollen records of late Quaternary North American sediments: American Association of Stratigraphic Palynologists Foundation, Dallas, p. 353–384.

Anderson, P. M., and Brubaker, L. R., 1986, Modern pollen assemblages from northern Alaska: Review of Palaeobotany and Palynology, v. 46, p. 273–291.

Ashworth, A. C., 1977, A late Wisconsinan Coleopterus assemblage from southern Ontario and its environmental significance: Canadian Journal of Earth Sciences, v. 14, p. 1625–1634.

Baker, B. H., and Kemperman, J. A., 1974, Spruce beetle effects on a white spruce stand in Alaska: Journal of Forestry, v. 72, p. 423–425.

Barnola, J. M., Raynaud, D., Korotkevich, Y. S., and Lorius, C., 1987, Vostok ice core provides 160,000-year record of atmospheric CO_2: Nature, v. 329, p. 408–414.

Bartlein, P. J., and Prentice, I. C., 1989, Orbital variations, climate and paleoecology: Trends in ecology and evolution, v. 4, p. 195–199.

Begét, J. E., 1990, Middle Wisconsin climatic fluctuations recorded in central Alaskan loess: Geographie Physique et Quaternaire, v. 44, p. 3–13.

Begét, J. E., and Hawkins, D. B., 1989, Influence of orbital parameters on Pleistocene loess deposition in central Alaska: Nature, v. 337, p. 151–153.

Begét, J. E., Stone, D. B., and Hawkins, D. B., 1990, Paleomagnetic forcing of magnetic susceptibility in Alaskan loess during the late Quaternary: Geology, v. 18, p. 40–43.

Begét, J. E., Edwards, M. E., Hopkins, D. M., Keskinana, M., and Kukla, G. J., 1991, Old Crow tephra found at the Palisades of the Yukon, Alaska: Quaternary Research, v. 35, p. 291–297.

Berger, G. W., 1985, Thermoluminescence dating studies of rapidly deposited silts from south-central British Columbia: Canadian Journal of Earth Sciences, v. 22, p. 704–710.

Berger, G. W., 1987, Thermoluminescence dating of Pleistocene Old Crow tephra and adjacent loess, near Fairbanks, Alaska: Canadian Journal of Earth Sciences, v. 24, p. 1975–1984.

Berger, G. W., 1988, Dating Quaternary events by luminescence, *in* Easterbrook, D. J., ed., Dating Quaternary sediments: Geological Society of America Special Paper 227, p. 13–50.

Berger, G. W., 1989, Comment on the status of TL dating of loess, *in* Carter, L. D., Hamilton, T. D., and Galloway, J. P., eds., Late Cenozoic history of the interior basins of Alaska and the Yukon: U.S. Geological Survey Circular 1026, p. 81–83.

Berger, G. W., 1990, Effectiveness of natural zeroing of the thermoluminescence in sediments: Journal of Geophysical Research, v. 95, p. 12375–12397.

Berger, G. W., 1991, The use of glass for dating volcanic ash by thermoluminescence: Journal of Geophysical Research, v. 96, p. 19705–19720.

Berger, G. W., 1994, Thermoluminescence dating of sediments older than 100 ka: Quaternary Science Reviews, v. 13, p. 445–455.

Berger, G. W. and Mahaney, W. C., 1990, Test of thermoluminescence dating of buried soils from Mt. Kenya, Kenya: Sedimentary Geology, v. 66, p. 45–56.

Berger, G. W. and Péwé, T. L., 1994, Chronology of interior Alaskan loess-paleosol deposits by thermoluminescence, *in* Lanphere, M. A., Dalrymple, G. B., and Turrin, B. D., eds., Abstracts: Eighth International Conference on Geochronology, Cosmochronology and Isotope Geology, U.S. Geological Survey Circular 1107, p. 28.

Berger, G. W., Burke, R. M., Carver, G. A., and Easterbrook, D. J., 1991, Test of thermoluminescence dating with coastal sediments from northern California: Isotope Geoscience, v. 87, p. 21–37.

Berger, G. W., Pillans, B. J., and Palmer, A. S., 1992, Dating loess up to 800 ka by thermoluminescence dating of loess from New Zealand and Alaska: Geology, v. 20, p. 403–406.

Berger, G. W., Pillans, B. J., and Palmer, A. S., 1994, Test of thermoluminescence dating of loess from New Zealand and Alaska: Quaternary Science Reviews, v. 13, p. 309–333.

Berger, G. W., Péwé, T. L., Westgate, J. A., and Preece, S. J., 1996, Age of Sheep Creek tephra (Pleistocene) in central Alaska from thermoluminescence dating of bracketing loess: Quaternary Research, v. 45, p. 263–270.

Blackwell, J. M., 1965, Surficial geology and geomorphology of the Harding Lake area, Big Delta Quadrangle, Alaska [M.S. thesis]: College, University of Alaska, 91 p.

Blasing, T. J. and Fritts, H. C., 1975, Past climate of Alaska and northwestern Canada as reconstructed from tree rings, *in* Weller, G., and Bowling, S. A., eds., Climate of the Arctic: Proceedings, Alaska Science Conference, 24th, August 1973: Fairbanks, Geophysical Institute, University of Alaska, p. 48–53.

Bloom, A. L., Broecker, W. S., Chappell, M. A., Matthews, R. K., and Mesolella, K. J., 1974, Quaternary sea-level fluctuations on a tectonic coast: New $^{230}Th/^{234}U$ data from the Huon Peninsula, New Guinea: Quaternary Research, v. 4, p. 185–205.

Bostock, H. S., 1948, Physiography of the Canadian Cordillera, with special reference to the area north of the fifty-fifth parallel: Geological Survey of Canada Memoir 247, 106 p.

Bostock, H. S., 1966, Notes on glaciation in central Yukon Territory: Geological Survey of Canada Paper 65-36, 14 p.

Bright, D. E., 1976, The insects and arachnids of Canada; the bark beetles of Canada and Alaska; Part 2: Ottawa, Canadian Department of Agriculture Publication 1576, 241 p.

Canfield, H. E., 1985, Thermoluminescence dating and chronology of loess deposition in the central United States [M.S. thesis]: Madison, University of Wisconsin, 15 p.

Capps, S. R., 1932, Glaciation in Alaska: U.S. Geological Survey Professional Paper 170-A, p. 1–8.

Carter, L. D., and Ager, T. A., 1989, Late Pleistocene spruce (Picea) in northern interior basins of Alaska and Yukon: Evidence from marine deposits in northern Alaska, *in* Carter, L. D., Hamilton, T. D., and Galloway, J. P., eds., Late Cenozoic history of interior basins of Alaska and the Yukon: U.S. Geological Survey Circular 1026, p. 11–14.

Chaney, R. W., and Mason, H. L., 1936, A Pleistocene flora from Fairbanks, Alaska: American Museum Novitates, v. 887, 17 p.

Chen, J. H., Curran, H. A., White, B., and Wassenberg, G. J., 1991, Precise chronology of the last interglacial period: ^{234}U-^{230}Th data from fossil coral reefs in the Bahamas: Geological Society of America Bulletin, v. 103, p. 82–97.

Cook, E. R., and Kariukstis, L. A., 1990, Methods of Dendrochronology: Applications in the Natural Sciences: Dordrecht, The Netherlands, Kluwer Academic Publishers, 394 p.

Coulter, H. W., Hopkins, D. M., Karlstrom, T. N. V., Péwé, T. L., Wahrhaftig, C., and Williams, J. R., 1965, Extent of glaciations in Alaska: U.S. Geological Survey Miscellaneous Geological Investigation Map I-415, scale 1:2,500,000.

Cropper, J. P., 1982, Climate reconstructions (1801 to 1938) inferred from tree-ring width chronologies of the North American arctic: Arctic and Alpine Research, v. 14, p. 223–241.

Cropper, J. P., and Fritts, H. C., 1981, Tree-ring width chronologies from the North American arctic: Arctic and Alpine Research, v. 13, p. 245–260.

D'Arrigo, R. D., Jacoby, J. C., and Free, R. M., 1992, Tree-ring width and maximum latewood density at the North American tree line: Parameters of climatic change: Canadian Journal of Forest Research, v. 22, p. 1290–1296.

Douglass, A. E., 1936, Climatic cycles and tree growth, Volume III: A study of cycles: Carnegie Institute of Washington Publication no. 289, 171 p.

Douglass, A. E., 1940, Examples of spiral compression wood: Tree-Ring Bulletin, v. 6, p. 21–22.

Edwards, M. E., and McDowell, P. F., 1989, Quaternary deposits at Birch Creek, northeastern interior Alaska: The possibility of climatic reconstruction, *in* Carter, L. D., Hamilton, T. D., and Galloway, J. P., eds., Late Cenozoic history of the interior basins of Alaska and Yukon: U.S. Geological Survey Circular 1026, p. 48–50.

Edwards, M. E., and McDowell, P. F., 1991, Interglacial deposits at Birch Creek, northeast interior Alaska: Quaternary Research, v. 35, p. 41–52.

Elias, S. A., and Edwards, M. E., 1982, Paleoenvironmental interpretation of bark beetle fossils from two high altitude sites in the Colorado Rockies: Proceedings, Third North American Paleontological Convention, Volume 1, p. 153–157.

Elias, S. A., Short, S. K., Nelson, C. H., and Birks, H. H., 1996, Life and times of the Bering land bridge: Nature, v. 382, p. 60–63.

Emiliani, C., 1955, Pleistocene temperatures: Journal of Geology, v. 63, p. 538–578.

Farquhar, G. D., O'Leary, M. H., and Baxter, J. A., 1982, On the relationship between carbon isotope discrimination and intercellular carbon dioxide concentration in leaves: Australian Journal of Plant Physiology, v. 9, p. 121–137.

Francey, R. J., and Farquhar, G. D., 1982, An explanation of $^{13}C/^{12}C$ variations in tree rings: Nature, v. 297, p. 28–31.

Fritts, H. C., 1969, Bristlecone pine in the White Mountains of California: Growth and ring width characteristics: Tucson, Papers of the Laboratory of Tree-Ring Research 4, University of Arizona Press, 44 p.

Fritts, H. C., 1976, Tree rings and climate: London, United Kingdom, Academic Press, 567 p.

Fritts, H. C., and Shatz, D. J., 1975, Selecting and characterizing tree-ring chronologies for dendroclimatic analysis: Tree-Ring Bulletin, v. 35, p. 31–40.

Fritts, H. C., and Swetnam, T. W., 1989, Dendroecology: A tool for evaluating variations in past and present forest environments: Advances in Ecological Research, v. 19, p. 111–188.

Geist, O. W., 1953, Collecting Pleistocene fossils in Alaska: Proceedings, Second Alaskan Science Conference, Mt. McKinley National Park, 1951, p. 171–172.

Giddings, J. L., Jr., 1938, Buried wood from Fairbanks, Alaska: Tree-Ring Bulletin, v. 4, p. 3–6.

Guthrie, R. D., 1968a, Paleoecology of the large-mammal community in interior Alaska: American Midland Naturalist: v. 79, p. 346–463.

Guthrie, R. D., 1968b, Paleoecology of late Pleistocene small mammal community from interior Alaska: Arctic, v. 22, p. 213–224.

Guthrie, R. D., 1982, Mammals of the Mammoth Steppe as paleoenvironmental indicators, *in* Hopkins, D. M., Matthews, J. V., Jr., Schweger, C. E., and Young, S. B., eds., Paleoecology of Beringia: New York, Academic Press, p. 307–329.

Guthrie, R. D., 1985, Woolly arguments against the Mammoth Steppe: A new look at the palynological data: Quaternary Review of Archaeology, v. 1, p. 9–16.

Guthrie, R. D., 1990, Frozen fauna of the Mammoth Steppe: The story of Blue Babe: Chicago, Illinois, University of Chicago Press, 323 p.

Hakala, J. B., 1952, The life history and general ecology of the beaver in interior Alaska [M.S. thesis]: College, University of Alaska, 189 p.

Hamilton, T. D., 1993, The Old Crow tephra: a stratigraphic marker for the last interglaciation in Alaska? *in* Kelmelis, J. A., and Snow, Mitchell, Proceedings, U.S. Geological Survey Global Change Research Forum, Herndon, Virginia, 1991: U.S. Geological Survey Circular 1086, p. 87.

Hamilton, T. D., and Brigham-Grette, J., 1991, The Last Interglaciation in Alaska: Stratigraphy and paleoecology of potential sites: Quaternary International, v. 10–12, p. 49–71.

Hamilton, T. D., Ager, T. A., and Robinson, S. W., 1983, Late Holocene ice wedges near Fairbanks, Alaska, U.S.A.: Environmental setting and history of growth: Arctic and Alpine Research, v. 15, p. 157–168.

Hansen, H. P., 1953, Postglacial forests in the Yukon Territory and Alaska: American Journal of Science, v. 251, p. 505–542.

Heusser, L. E. and King, J. E., 1988, North America, *in* Huntley, B., and Webb, T., III, eds., Vegetation history: Dordrecht, The Netherlands, Kluwer Academic Publishers, p. 193–236.

Holmes, G. W., and Péwé, T. L., 1965, Geological map of the Mount Hayes D-3 Quadrangle, Alaska: U.S. Geological Survey Geological Quadrangle Map GQ-366, scale 1:63,160, 1 sheet.

Holmes, R. L., 1983, Computer-assisted quality control in tree-ring dating and measurement: Tree-Ring Bulletin, v. 44, p. 69–75.

Hopkins, D. M., 1972, The paleogeography of Beringia during late Cenozoic: Internord, v. 12, p. 121–150.

Hopkins, D. M., 1982, Aspects of the paleogeography of Beringia during the late Pleistocene, *in* Hopkins, D. M., Matthews, J. V., Jr., Schweger, C. E., and Young, S. B., eds., Paleoecology of Beringia: New York, Academic Press, p. 3–28.

Hughes, M. K., Kelly, P. M., Pilcher, J. R., and LaMarche, V. C., Jr., eds., 1982, Climate from tree rings: Cambridge, United Kingdom, Cambridge University Press, 223 p.

Imbrie, J., and eight others, 1984, The orbital theory of Pleistocene climate: Support from a revised chronology of the marine $\delta^{18}O$ record, *in* Berger, A. L., Imbrie, J., Hays, J., Kukla, G., and Sal, B., eds., Milankovitch and climate, Part 1: Boston, Massachusetts, D. Reidel Publishing Co., p. 269–305.

Jouzel, J., and 16 others, 1993, Extending the Vostok ice core record of palaeoclimate to the penultimate glacial period: Nature, v. 364, p. 407–412.

Krishnamurthy, R. V., and Epstein, S., 1990, Glacial-interglacial excursion in the concentration of atmospheric CO_2: Effect in the $^{13}C/^{12}C$ ratio in wood cellulose: Tellus, v. 42B, p. 423–434.

Ku, T. L., Kimmel, M. A., Easton, W. H., and O'Neil, T. J., 1974, Eustatic sea-level 120,000 years ago on Oahu, Hawaii: Science, v. 183, p. 959–912.

Larsen, J. A., 1980, The Boreal Ecosystem: New York, Academic Press, 500 p.

Leavitt, S. W., 1993, Environmental information from $^{13}C/^{12}C$ ratios of wood: in Climate Change in Continental Isotope Records, Geophysical Monograph 78, American Geophysical Union, Washington, D.C., p. 325–331.

Leavitt, S. W., and Danzer, S. R., 1991, Chronology from plant matter: Nature, v. 352, p. 671.

Leavitt, S. W. and Danzer, S. R., 1992, $\delta^{13}C$ variations in C_3 plants over the past 50,000 years: Radiocarbon, v. 34, p. 783–791.

Leavitt, S. W., and Danzer, S. R., 1993, Method for batch processing small wood samples to holocellulose for stable-carbon isotope analysis: Analytical Chemistry, v. 65, p. 87–89.

Leavitt, S. W., and Long, A., 1986, Stable-carbon isotope variability in tree foliage and wood: Ecology, v. 67, p. 1002–1010.

Leavitt, S. W., and Long, A., 1988, Stable carbon isotope chronologies from trees in the southwestern United States: Global Biogeochemical Cycles, v. 1, p. 189–198.

Leavitt, S. W., and Long, A., 1989, Drought indicated in carbon-13/carbon-12 ratios of southwestern tree rings: Water Resources Bulletin, v. 25, p. 341–347.

Leavitt, S. W., and Newberry, T., 1992, Systematics of stable-carbon isotopic differences between gymnosperm and angiosperm trees: Plant Physiology (Life Science Advances), v. 11, p. 257–262.

Leuenberger, M., Siegenthaler, U., and Langway, C. C., 1992, Carbon isotope composition of atmospheric CO_2 during the last ice age from an Antarctic ice core: Nature, v. 357, p. 488–490.

Libby, L. M., Pandolfi, L. F., Payton, P. H., Marshall, J., Becker, B., and Gierte-Siebenlist, V., 1976, Isotope tree thermometers: Nature, v. 261, p. 284–288.

Lichti-Federovich, S., 1973, Palynology of six sections of late Quaternary sediments from the Old Crow River, Yukon Territory, Canada: Canadian Journal of Botany, v. 51, p. 553–564.

Long, A. and Kalin, R. M., 1992a, High-sensitivity radiocarbon dating in the 50,000 to 70,000 BP range without isotopic enrichment: Radiocarbon, v. 34, p. 351–359.

Long, A., and Kalin, R. M., 1992b, Radiocarbon dating in the 50,000 to 65,000 year range without isotopic enrichment, *in* Povinec, P., ed., Proceedings, 14th Europhysics Conference on Nuclear Physics: Rare Nuclear Processes: Singapore, Word Scientific, p. 256–263.

Lutz, H. J., 1956, Ecological effects of forest fires on the vegetation of interior Alaska: U.S. Forest Service Technical Bulletin 1133, 121 p.

Marino, B. D., McElroy, M. B., Salawitch, R. J., and Spaulding, W. G., 1992, Glacial-to-interglacial variations in the carbon isotopic composition of atmospheric CO_2: Nature, v. 357, p. 461–466.

Martinson, D. G., Pisias, N. G., Hays, J. D., Imbrie, J., Moore, T. C., Jr., and Shackleton, N. D., 1987, Age dating and the orbital theory of the ice ages: Development of a high-resolution 0 to 300,000 year chronostratigraphy: Quaternary Research, v. 27, p. 1–30.

Matthews, J. V., Jr., 1968, A paleoenvironment analyses of three late Pleistocene Coleopterous assemblages from Fairbanks, Alaska: Quaestiones entomologicae, v. 4, p. 202–224.

Matthews, J. V., Jr., 1970, Quaternary environmental history of interior Alaska; pollen samples from organic colluvium and peats: Arctic and Alpine Research, v. 2, p. 241–251.

Matthews, J. V., Jr., 1974, Wisconsin environment of interior Alaska: Pollen and macrofossil analyses of 27 m core from Isabella Basin (Fairbanks, Alaska): Canadian Journal of Earth Sciences, v. 11, p. 812–841.

Matthews, J. V., Jr., Hughes, O. L., and Schweger, C. E., 1987, Twelve mile Bluff exposure, *in* Morison, S. R., and Smith, C. A. S., eds., Guidebook to Quaternary research in Yukon: International Union of Quaternary Research, XII Congress: Ottawa, Ontario, Canada, National Research Council of Canada, p. 95–99.

Matthews, J. V., Jr., Schweger, C. E., and Hughes, O. L., 1989, Climatic change in eastern Beringia during Oxygen Isotope Stages 2 and 3: Proposed thermal events, *in* Carter, L. D., Hamilton, T. D., and Galloway, J. P., eds., Late Cenozoic history of interior basins of Alaska and the Yukon: U.S. Geological Survey Circular 1026, p. 34–38.

Matthews, J. V., Jr., Schweger, C. E., and Janssens, J. A., 1990, The last (Koy-Yukon) interglaciation in the northern Yukon: Evidence from unit 4 at Ch'ijee's Bluff, Bluefish Basin: Geographie Physique et Quaternaire, v. 44, p. 341–362.

McCormac, F. G., Kalin, R. M., and Long, A., 1993, Radiocarbon dating beyond 50,000 years by liquid scintillation counting, *in* Proceedings, International Conference on advances in liquid scintillation spectrometry: Vienna, Austria, p. 1–9.

McKay, E. D., 1986, Illinoian and older loesses and tills at the Maryville Section: Illinois State Geological Survey American Quaternary Association Guidebook, p. 21–30.

Morgan, A. V., and Morgan, A., 1979, The fossil Coleoptera of the Two Creeks Forest Bed, Wisconsin: Quaternary Research, v. 12, p. 226–240.

Muhs, D. R., and Szabo, B. J., 1991, New uranium-series ages of the Waimanalo Limestone, Oahu, Hawaii and paleoclimatic implications for the last interglacial period: Geological Society of America Abstracts with Programs, v. 23, no. 5, p. A239.

Murray, D. F., 1961, Some factors affecting the production and harvest of beaver in the upper Tanana River, Alaska [M.S. thesis]: College, University of Alaska, 103 p.

Naeser, N. D., Westgate, J. A., Hughes, O. L. and Péwé, T. L., 1982. Fission-track ages of late Cenozoic distal tephra beds in the Yukon Territory and Alaska: Canadian Journal of Earth Sciences, v. 19, p. 2167–2178.

NOAA [U.S. National Oceanic and Atmospheric Administration], 1989, Local climatological data, Fairbanks, Alaska: 1988 annual summary: Asheville, North Carolina, NOAA, 4 p.

Norton, L. D., and Bradford, J. M., 1985, Thermoluminescence dating of loess from western Iowa: Soil Science Society of America Journal, v. 49, p. 708–712.

Petit, J. R., Briat, M., and Royer, A., 1981, Ice age aerosol content from East Antarctic ice core samples and past wind strength: Nature, v. 293, p. 391–394.

Péwé, T. L., 1950, Origin and upland silt in the Fairbanks area, Alaska [abs.]: Geological Society of America Bulletin, v. 61, p. 1493.

Péwé, T. L., 1951, An observation on wind-blown silt: Journal of Geology, v. 59, p. 399–401.

Péwé, T. L., 1952a, Geomorphology of the Fairbanks area, Alaska [Ph.D. dissert.]: Stanford, California, Stanford University, 220 p.

Péwé, T. L., 1952b, Preliminary report of late Quaternary history of the Fairbanks area, Alaska [abs.]: Geological Society of America Bulletin, v. 63, p. 1289–1290.

Péwé, T. L., 1952c, Preliminary report of multiple glaciation in the Big Delta area, Alaska [abs.]: Geological Society of America Bulletin, v. 63, p. 1289.

Péwé, T. L., 1954, Effect of permafrost on cultivated fields near Fairbanks, Alaska: U.S. Geological Survey Bulletin 989-F, p. 315–351.

Péwé, T. L., 1955, Origin of the upland silt near Fairbanks, Alaska: Geological Society of America Bulletin, v. 66, p. 699–724.

Péwé, T. L., 1957, Permafrost and its effect on life in the North: Arctic Biology: Corvallis, 18th Biological Colloquium, Oregon State College, p. 12–25.

Péwé, T. L., 1958, Geology of the Fairbanks D-2 quadrangle, Alaska: U.S. Geological Survey Geologic Quadrangle Map GQ-110, scale 1:63,360, 1 sheet.

Péwé, T. L., 1962, Ice wedges in permafrost, lower Yukon River area near Galena, Alaska: Bulletin Peryglacjalny, Series, v. 3, p. 65–76.

Péwé, T. L., 1965a, Delta River area, Alaska Range, *in* Péwé, T. L., Ferrians, O. J., Jr., Nichols, V. R., and Kalstrom, T. N. V., Guidebook for Field Conference F, Central and South-Central Alaska: International Union for Quaternary Research, 7th Congress: Lincoln, Nebraska Academy of Science, p. 55–93. (Slightly modified version reprinted 1977 by Alaska Division of Geological and Geophysical Surveys.)

Péwé, T. L., 1965b, Middle Tanana River Valley, *in* Péwé, T. L., Ferrians, O. J., Jr., Karlstrom, T. N. V., and Nichols, D. R., Guidebook for Field Conference F, Central and South-Central Alaska, International Union for Quaternary Research, 7th Congress: Lincoln, Nebraska Academy of Science, p. 36–54. (Slightly modified version reprinted 1977 by Alaska Division of Geological and Geophysical Surveys.)

Péwé, T. L., 1965c, Fairbanks area, *in* Péwé, T. L., Ferrians, O. J., Jr., Nichols, D. R., and Karlstrom, T. N. V., Guidebook for Field Conference F, Central and South-Central Alaska—International Union for Quaternary Research, 7th Congress, U.S.A.: Lincoln, Nebraska, Nebraska Academy of Sciences, p. 6–36. (Slightly modified version reprinted in 1977 by Alaska Division of Geological and Geophysical Surveys.)

Péwé, T. L., 1966, Ice wedges in Alaska—classification, distribution, and climatic significance, *in* Proceedings, International Permafrost Congress, Lafayette, Indiana, 1963: National Academy of Sciences–National Research Council Publications, no. 1287, p. 76–81.

Péwé, T. L., 1975a, Quaternary geology of Alaska: U.S. Geological Survey Professional Paper 835, 145 p.

Péwé, T. L., 1975b, Quaternary stratigraphic nomenclature in central Alaska: U.S. Geological Survey, Professional Paper 862, 32 p.

Péwé, T. L., 1982, Geologic hazards of the Fairbanks area, Alaska: Alaska Division of Geological and Geophysical Surveys Special Report 15, 109 p.

Péwé, T. L., 1989, Quaternary stratigraphy of the Fairbanks area, Alaska, *in* Carter, L. D., Hamilton, T. D., and Galloway, J. P., eds., Late-Cenozoic history of the interior basins of Alaska and the Yukon: U.S. Geological Survey Circular 1026, p. 72–77.

Péwé, T. L., 1992, Origin and age of erosional gullies and ridges in upper hillside loess of central Alaska—evidence of ancient global warming: Geological Society of America Abstracts and Programs, v. 24, no. 7, p. A51.

Péwé, T. L., in press, Late Cenozoic stratigraphy of the Fairbanks area in eastern Beringia: An informal status report, *in* Edwards, M. E., and Sher, A., eds., Common problems in terrestrial paleoenvironmental studies in Beringia: Fairbanks, University of Alaska Press.

Péwé, T. L., and Bell, J. W., 1974, Map showing distribution of permafrost in the Fairbanks D-2SW Quadrangle, Alaska: U.S. Geological Survey, Miscellaneous Investigation Series, I-829-B, scale 1:24,000, 1 sheet.

Péwé, T. L., and Holmes, G. W., 1964, Geology of the Mount Hayes D-4 Quadrangle, Alaska: U.S. Geological Survey Miscellaneous Investigation Map I-394, scale 1:63,360, 2 sheets.

Péwé, T. L., and Hopkins, D. M., 1967, Mammal remains of pre-Wisconsin age in Alaska, *in* Hopkins, D. M., ed., The Bering Land Bridge: Stanford, California, Stanford University Press, p. 266–270.

Péwé, T. L., and Paige, R. A., 1963, Frost heaving of piles with an example from Fairbanks area, Alaska: U.S. Geological Survey Bulletin 1111-I, p. 333–407.

Péwé, T. L., and Reger, R. D., 1983, Guidebook to permafrost and Quaternary geology along the Richardson and Glenn Highways between Fairbanks and Anchorage, Alaska, *in* International Conference on Permafrost, 4th, Alaska Division of Geological and Geophysical Surveys, Guidebook 1, 263 p.

Péwé, T. L., and Reger, R. D., 1989, Quaternary geology and permafrost along the Richardson and Glenn Highways between Fairbanks and Anchorage, Alaska, *in* International Geological Congress, 28th, Fieldtrip Guidebook,

T102: Washington, D.C., American Geophysical Union, 54 p.

Péwé, T. L., and Sellmann, P. V., 1973, Geochemistry of permafrost and Quaternary stratigraphy, *in* Permafrost: North American contribution to 2nd International Conference: Washington, D.C., National Academy of Sciences, p. 166–170.

Péwé, T. L., and eight others, 1953, Multiple glaciation in Alaska—A progress report: U.S. Geological Survey Circular 289, 13 p.

Péwé, T. L., Wahrhaftig, C., and Weber, F. R., 1966, Geologic map of the Fairbanks Quadrangle, Alaska: U.S. Geological Survey Miscellaneous Investigations Map I-455, scale 1:250,000.

Péwé. T. L., Burbank, L., and Mayo, L. R., 1967, Multiple glaciation in the Yukon-Tanana Upland, Alaska: U.S. Geological Survey Miscellaneous Investigations Map I-507, scale 1:500,000.

Péwé, T. L., Reger, R. D., and Westgate, J. A., 1989, Fairbanks area, *in* Péwé, T. L., and Reger, R. D., Quaternary geology and permafrost along the Richardson and Glenn Highways between Fairbanks and Anchorage, Alaska, *in* International Geological Congress, 28th, Fieldtrip Guidebook, T102: Washington, D.C., American Geophysical Union, p. 3–16.

Péwé, T. L., Berger, G. W., and Westgate, J. A., 1995, Past major global warming in 3 Myr loess record, central Alaska, USA, indicated by three buried taiga forest beds: 2 Myr, 125 ka, 10 ka: Berlin, Germany, International Union for Quaternary Research (Abstract Volume), p. 217.

Porter, L., 1988, Late Pleistocene fauna of Lost Chicken Creek, Alaska: Arctic, v. 41, p. 303–313.

Preece, S. J., 1991, Tephrostratigraphy of the late Cenozoic Gold Hill Loess, Fairbanks, Alaska [M.S. thesis]: Toronto, Ontario, University of Toronto, 164 p.

Repenning, C. A., Hopkins, D. M., and Rubin, M., 1964, Tundra rodents in a late Pleistocene fauna from the Tofty district, central Alaska: Arctic, v. 17, p. 177–197.

Robinson, W. J., and Evans, R., 1980, A microcomputer-based tree-ring measuring system: Tree-Ring Bulletin, v. 40, p. 59–64.

SAS Institute, Inc., 1986, SAS System for Linear Models, 1986 Edition: Cary, North Carolina 329 p.

Schweger, C. E., 1989, The Old Crow and Bluefish Basins, Northern Yukon: Development of the Quaternary history, *in* Carter, L. D., Hamilton, T. D., and Galloway, J. P., eds., Late Cenozoic history of interior basins of Alaska and the Yukon: U.S. Geological Survey Circular 1026, p. 30–33.

Schweger, C. E., and Matthews, J. V., Jr., 1985, Early and middle Wisconsinan environments of eastern Beringia: Stratigraphic and paleoecology implications of the Old Crow tephra: Geographie Physique et Quaternaire, v. 39, p. 275–290.

Schweingruber, F. H., 1987, Tree rings: Basics and applications of dendrochronology: Dordrecht, The Netherlands, D. Reidel, 276 p.

Schweingruber, F. H., Fritts, H. C., Braker, O. U., Drew, L. G., and Schär, E., 1978, The X-ray technique as applied to dendroclimatology: Tree-Ring Bulletin, v. 38, p. 61–91.

Sherman, C. E., Glenn, C. R., Jones, A. T., Burnett, W. C., and Schwarcz, H. P., 1993, New evidence for two highstands of the sea during the last interglacial, oxygen isotope substage 5e: Geology, v. 21, p. 1079–1082.

Shostakovitch, V. B., 1925, Forest conflagrations in Siberia, with special references to the fires of 1915: Journal of Forestry, v. 23, p. 365–371.

Stuiver, M., 1978, Atmospheric carbon dioxide and carbon reservoir changes: Science, v. 199, p. 253–258.

Tarnocai, C., and Schweger, C. E., 1991, Late Tertiary and early Pleistocene paleosols in northwestern Canada: Arctic, v. 44, p. 1–11.

Telewski, F. W., 1988, Intra-annual spiral compression wood: A record of low-frequency gravitropic circumnutational movement in trees: International Association of Wood Anatomy Bulletin, v. 9, p. 269–274.

Thompson, L. G., and Thompson-Mosley, E., 1981a, Micropartical concentration variations linked with climatic change: Evidence from polar ice cores: Science, v. 212, p. 812–815.

Thompson, L. G., and Thompson-Mosley, E., 1981b, Temporal variability of microparticle properties in polar ice sheets: Journal of Volcanology and Geothermal Research, v. 11, p. 11–27.

U.S. Weather Bureau, 1943, Climatic atlas for Alaska: Army Air Forces Weather Information Branch Headquarters, Report 44, 229 p.; Supplement, 72 p.

Van de Water, P. K., Leavitt, S. W., and Betancourt, J. L., 1994, Trends in stomatal density and $^{13}C/^{12}C$ ratios of *Pinus flexilis* needles during last glacial-interglacial cycle: Science, v. 264, p. 239–264.

Viereck, L. A., 1973, Wildfire in the taiga of Alaska: Quaternary Research , v. 3, p. 465–495.

Viereck, L. A., 1975, Forest ecology of the Alaska taiga, *in* Proceedings, Circumpolar Conference on Northern Ecology, Ottawa: Natural Research Council of Canada, p. I1–I22.

Viereck, L. A., and Little, E. L., Jr., 1972, Alaska trees and shrubs: Washington, D.C., U.S. Department of Agriculture Handbook 410, 265 p.

Wagner, G. A., and Van den haute, P., 1992, Fission-track dating: Stuttgart, Germany, Ferdinand Enke Verling, 274 p.

Watson, C. E., 1959, Climate of Alaska, *in* Climate of the states: U.S. Weather Bureau, Climatology of the United States, no. 60-49, 24 p.

Watts, W. A., 1988, Europe, *in* Huntley, B., and Webb, T., III, eds., Vegetation history: Dordrecht, The Netherlands, Kluwer Academic Publishers, p. 155–191.

Weber, F. R., 1986 Glacial geology of the Yukon-Tanana Upland, *in* Hamilton, T. D., Reed, K. M., and Thorson, R. M., eds., Glaciation in Alaska; the geological record: Anchorage Geological Society, p. 74–98.

Weber, F. R., Hamilton, T. D., Hopkins, D. M., Repenning, C. A., and Haas, H., 1981, Canyon Creek: A late Pleistocene vertebrate locality in interior Alaska: Quaternary Research, v. 16, p. 167–180.

Werner, R. A., and Helsten, E. H., 1984, Scolytidae associated with felled white spruce in Alaska: The Canadian Entomologist, v. 116, p. 465–471.

Werner, R. A., Baker, B. H., and Rush, P. A., 1977, The spruce beetle in white spruce forests of Alaska: U.S. Department of Agriculture Technical Report PNW-61, 13 p.

Westgate, J. A., 1987, Compositional comparisons of Old Crow tephra, Sheep Creek tephra, and White River tephra and its significance for the provenence of the wide-spread Beringian tephras: Ottawa, Ontario, Canada, International Union for Quaternary Research, Abstracts, p. 288.

Westgate, J. A., 1988, Isothermal plateau fission-track age of the late Pleistocene Old Crow tephra, Alaska: Geophysical Research Letters, v. 15, p. 376–379.

Westgate, J. A., 1989, Isothermal plateau fission-track age of hydrated glass shards from silicic tephra beds: Earth and Planetary Science Letters, v. 95, p. 226–234.

Westgate, J. A., Hamilton, T. D., and Gorton, M. P., 1983, Old Crow tephra: A new Late Pleistocene stratigraphic marker across north-central Alaska and western Yukon Territory: Quaternary Research, v. 19, p. 39–54.

Westgate, J. A., Walter, R. C., Pearce, G. W., and Gorton, M. P., 1985, Distribution, stratigraphy, geochemistry, and palaeomagnetism of the late Pleistocene Old Crow tephra in Alaska and the Yukon: Canadian Journal of Earth Sciences, v. 22, p. 893–906.

Westgate, J. A., Stemper, B. A., and Péwé, T. L., 1990, A 3 m.y. record of Pliocene-Pleistocene loess in interior Alaska: Geology, v. 18, p. 858–861.

Winograd, I. J., Coplen, T. B., Ludwig, K. R., Szabo, B. J., Riggs, A. C., and Revesz, K. M., 1990, Continuous 500,000 year climate record from Great Basin vein calcite: 1. The $\delta^{18}O$ time series: Geological Society of America Abstracts with Programs, v. 22, no. 7, p. A209.

Winograd, I. J., Coplen, T. B., Ludwig, K. R., Szabo, B. J., Riggs, A. C., and Revesz, K. M., 1992, Continuous 500,000 year climate record from vein calcite in Devil's Hole, Nevada: Science, v. 258, p. 255–260.

Wintle, A. G., and Huntley, D. J., 1980, Thermoluminescence dating of ocean sediments: Canadian Journal of Earth Sciences, v. 17, p. 348–360.

Wintle, A. G., and Westgate, J. A., 1986, Thermoluminescence age of Old Crow tephra in Alaska, Geology, v. 14, p. 594–597.

Yeend, W. E., 1977, Tertiary and Quaternary deposits at the Palisades, central Alaska: Journal of Research of the U.S. Geological Survey, v. 5, p. 747–752.

MANUSCRIPT ACCEPTED BY THE SOCIETY JANUARY 28, 1997

Printed in U.S.A.

Typeset and printed in U.S.A. by Johnson Printing, Boulder, Colorado